VISIBLE LIGHT COMMUNICATIONS

Edited by **Jin-Yuan Wang**

Visible Light Communications
http://dx.doi.org/10.5772/66915
Edited by Jin-Yuan Wang

Contributors

Peng Deng, Jian Dang, Mengting Wu, Liang Wu, Zaichen Zhang, Bugra Turan, Seyhan Ucar, Shuxu Guo, Tian Zhang, Jun Yao, Wasiu O. Popoola, Hammed G. Olanrewaju, Funmilayo B. Offiong, Ahmed Taha Hussein, Mohammed Alresheedi, Jaafar Elmirghani, Qi Wang, Tianqi Mao, Zhaocheng Wang

Published by InTech
Janeza Trdine 9, 51000 Rijeka, Croatia

Publishing Process Manager Martina Usljebrka
Technical Editor SPi Global
Cover InTech Design team

First published July, 2017
Printed in Croatia
Legal deposit, Croatia: National and University Library in Zagreb

Additional hard copies can be obtained from orders@intechopen.com

Visible Light Communications, Edited by Jin-Yuan Wang
p. cm.
Print ISBN 978-953-51-3407-7
Online ISBN 978-953-51-3408-4

For EU product safety concerns:
IN TECH d.o.o., Prolaz Marije Krucifikse Kozulić 3, 51000 Rijeka, Croatia,
info@intechopen.com or visit our website at intechopen.com.

Meet the editor

Jin-Yuan Wang received his PhD degree at the National Mobile Communications Research Laboratory, Southeast University, China, in 2015. He is now a lecturer at Peter Grünberg Research Center, Nanjing University of Posts and Telecommunications, Nanjing, China. His current research interest is visible light communications. He has authored/coauthored over 60 journal/conference papers. He was the Technical Program Committee member of ISWTA 2014 in Kota Kinabalu, ICCN 2014 in Wuhan IEEE ICC 2015 in London, and WTS 2017 in Chicago. He also serves as a reviewer for many journals, such as *IEEE Transactions on Wireless Communication*, *IEEE Transactions on Communications*, and *IEEE Transactions on Vehicular Technology and Optics Express*.

Contents

Preface

In the last decade, the field of visible light communication (VLC) has evolved rapidly. Compared to the traditional radio frequency wireless communications (RFWC), VLC has many advantages, such as worldwide availability, high security, large bandwidth, immunity to radio frequency interference, and unlicensed spectrum. Due to its superiority, VLC has become a complementary solution to the overcrowded RFWC.

Having seen the rapid developments of VLC in recent years, I feel that there is a need for a new textbook on this topic. This book intends to introduce the latest research progress in VLC, which covers four important aspects of VLC: (a) the novel modulation techniques for VLC, (b) the multiple-input multiple-output (MIMO) techniques for VLC, (c) the collaborative communication techniques for VLC, and (d) the practical applications of VLC. I have tried to make the book as useful as possible as a resource for researchers, engineers, scientists, and students interested in understanding and designing VLC systems.

The book is organized into seven chapters and describes various topics related to VLC. *Chapter 1* investigates a low-complexity modulation technique—spatial modulation for VLC. This chapter first introduces the principle of optical spatial modulation, and the error performance of optical spatial modulation is also analyzed. After that, comparison between optical spatial modulation and conventional modulation techniques is presented. In *Chapter 2*, a real-time software-defined adaptive MIMO VLC is experimentally demonstrated. This chapter shows the nonlinear modulation characteristics of the light-emitting diode in VLC, including nonlinear electrical-optical characteristics, nonlinear modulation bandwidth, and the effects of the nonlinear modulation on VLC performance. Then, the adaptive MIMO VLC is proposed, and the experimental setup and results are presented. *Chapters 3–5* focus on the optical orthogonal frequency-division multiplexing (OFDM) techniques. Specifically, *Chapter 3* designs the transceiver for direct current-biased optical (DCO)-OFDM in MIMO VLC. The model and the design for the conventional receiver in DCO-OFDM are first introduced. To suppress the clipping noise, advanced receiver design for DCO-OFDM is proposed. Then, the transceiver for MIMO DCO-OFDM is designed. *Chapter 4* proposes an index modulation-aided OFDM for VLC. The existing optical OFDM techniques are summarized. After that, the index modulation-aided OFDM for VLC is analyzed. The challenges and opportunities are discussed for the deployment of index modulation-aided OFDM in VLC systems. *Chapter 5* investigates the peak-to-average-power ratio (PAPR) reduction schemes for optical OFDM-based VLC. In this chapter, the Gaussian blur-based PAPR reduction scheme and the OFDM pulse-width modulation (PWM)-based PAPR reduction scheme are proposed to mitigate the PAPR. In *Chapter 6*, a collaborative VLC/infrared optical wireless (IROW) system is proposed. This chapter introduces the imaging VLC system, five IROW systems, and collaborative VLC/IROW system. Then, adaptive rate technique is proposed to evaluate the performance of the imaging VLC system. Finally, in *Chapter 7*, VLC is applied to the intelligent transportation system. The vehicular VLC (V2LC) is investigated. The transmitter, the receiver, and the channel for the V2LC are introduced. Then, the security implications of V2LC and the possible approaches to secure the vehicular communication are studied.

I would like to thank all contributors of the book, who have made it what it is; without them, the book would never have been realized. Many thanks are also directed to Prof. Ming Chen and Prof. Jun-Bo Wang at Southeast University for introducing me to the area and for fostering my early explorations in VLC. I also express my sincere gratitude to Prof. Yongjin Wang at Nanjing University of Posts and Telecommunications for his support. The book editor would like to acknowledge the open research fund of the National Mobile Communications Research Laboratory, Southeast University (No. 2017D06) and NUPTSF (No. NY216009), the open research fund for Jiangsu Key Laboratory for Traffic and Transportation Security (Huaiyin Institute of Technology) (No. TTS2017-03), and the open research fund of Key Laboratory of Intelligent Computing & Signal Processing (Anhui University) for their support of VLC-related research activities.

Finally, special thanks are extended to the publishing process manager Ms. Martina Usljebrka for her tremendous effort in organizing the logistics of the book, including the editing and promotion that allowed this book to happen.

Dr. Jin-Yuan Wang
Nanjing University of Posts and Telecommunications
Nanjing, China

Spatial Modulation – A Low Complexity Modulation Technique for Visible Light Communications

Hammed G. Olanrewaju,
Funmilayo B. Ogunkoya and Wasiu O. Popoola

Additional information is available at the end of the chapter

http://dx.doi.org/10.5772/intechopen.68888

Abstract

In visible light communication (VLC), the fundamental limitation on the achievable data rate/spectral efficiency is imposed by the optical source, particularly the phosphor-converted white light emitting diode (LED). These low-cost white LEDs favoured in solid-state lighting have very limited modulation bandwidth of less than 5 MHz, typically. This imposes a severe limitation on the attainable data rate. This is recognised in the literature and has led to the emergence of techniques such as multiple-input-multiple-output (MIMO) VLC systems as a means of addressing this challenge. The MIMO approach takes advantage of the multi-LED/multi-receiver structure to improve performance. In this chapter, we shall be discussing spatial modulation (SM) as a novel low-complexity MIMO technique for the VLC system. The SM technique exploits the spatial location of the individual LED as an additional degree of freedom in data modulation. Moreover, the chapter includes the comparison analysis of the SM technique with other traditional methods of modulation such as on-off keying (OOK) and pulse position modulation (PPM).

Keywords: spatial modulation, optical wireless communication, visible light communication, space shift keying

1. Introduction

The need for high data rate transmission to meet the demands of existing and emerging data-intensive applications and services is one of the key elements driving the research in the area of wireless communications access technology. Visible light communication (VLC) system, with its significantly large inherent optical spectrum, is a promising technology capable of delivering high data rates. However, one of its major drawbacks is the limited modulation bandwidth of light emitting diodes (LEDs). This limitation prevents some of the conventional wireless modulation techniques from fully exploiting the huge optical spectrum in achieving high-speed communications. To fully utilise the inherent resources of VLC while mitigating the effects of this limitation, a multiple-input multiple-output (MIMO) access technique such as spatial modulation (SM) is an attractive option.

SM technique in VLC facilitates efficient management of the limited LED modulation bandwidth in a power-efficient manner without sacrificing the complexity of the system. A study in Ref. [1] proposed SM for optical wireless systems using the same principle of SM for a radio frequency (RF) system [2]. One of the interesting features of VLC is the possibility of the system to serve a dual function of data transmission by intensity-modulating LEDs alongside their primary purpose of illumination. In practice, an LED luminaire usually has multiple LEDs used for illumination due to the limited luminous flux of an individual LED. Thus, SM leverages the spatial dimensions of these multiple LEDs as an additional degree of freedom for high data rate transmission [1]. The information sequence to be transmitted is mapped to symbols chosen from the signal constellation points of a digital modulation technique. The fundamental concept of SM relies on these spatially separated LEDs considered as spatial constellation points, which are utilised to convey additional information bits [3]. During a symbol duration, the transmit bits are mapped to one of the spatial constellation points, thereby activating one transmit unit only at a particular time instance. The strength of channel correlation between the transmit-receive unit plays a significant role in the performance of the SM technique. Hence, there is a need for intensive research on performance enhancement schemes of SM.

Various optical SM techniques in VLC reported in the literature to enhance the data rate include spatial pulse position modulation (SPPM) proposed in Ref. [4], where a combination of space shift keying (SSK) and pulse position modulation (PPM) is considered. The idea was later generalised in Ref. [5] by activating multiple LEDs during each symbol's duration in order to increase the number of bits transmitted at a time instance. The use of optical space shift keying (OSSK) [6], spatial pulse amplitude modulation (S-PAM) [7], and generalised space shift keying (GSSK) [8] to boost VLC spectral efficiency using SM has equally been investigated.

The focus of this chapter is to address the error performance challenges of SM in VLC. A detailed description of the conventional SM and different variants of SM is presented. Analytical expressions for the error performance evaluations of these variants are derived in the presence of VLC channel impairments and additive white Gaussian noise (AWGN). These solutions are compared with computer-based simulations to validate the closed-form expressions. Furthermore, the attainable data rate using SM is quantified by comparing the spectral efficiency of SM with the classical modulation techniques such as on-off keying (OOK) and PPM.

2. Principle of optical spatial modulation

2.1. Generation and detection

In a VLC system with N_t LED transmitting unit and N_r receivers (photodetectors (PD)), using optical SM technique, only one of the LEDs is active during any symbol duration while the rest of the LEDs is idle. For applications in VLC, an idle LED is driven by only the DC bias required to turn on the LED for illumination, while the active LED is driven by the sum of the DC bias and a current swing whose magnitude is dependent on the digital signal modulation imposed on the optical carrier. At the transmitter, the information bits to be transmitted are grouped into data symbols, and the block of information bits in each data symbol is mapped onto two information carrying units, namely spatial and signal constellations points. The total number of bits transmitted per symbol is given by $M = \log_2(LN_t)$, where L denotes the number of the digital signal constellation points. The spatial modulator at the transmitter maps the first $\log_2(N_t)$ most significant bits of each symbol to the spatial constellation point, which is used to select the LED that will be active, while the remaining $\log_2(L)$ bits are mapped to the signal constellation point. The signal constellation point is conveyed by the transmitted digital signal modulation such as pulse amplitude modulation (PAM) [9], PPM [4], and orthogonal frequency division multiplexing (OFDM) [10], among others. For instance, if PAM is used for signal modulation, the signal constellation point determines the intensity level of the optical signal that will be emitted by the active LED. As an illustration, the encoding mechanism for an optical SM scheme with $N_t = 4$, $L = 4$, and $M =$ 4 bits/symbol is shown in **Figure 1**. Considering the first data symbol which consists of bits '1101', LED 4 is selected to be activated based on the first two most significant bits, '11'. The remaining two bits, '01', are encoded in the signal constellation point 'S2'.

The signal emitted by the active LED propagates through the optical wireless channel. At the receiving end, the radiated signal is detected by one or more PDs. The channel condition of each transmitter-receiver path is described by the channel impulse response $h(t)$, and this is typically fixed for a specified configuration of the transmitter, receiver, and any intervening reflectors [11]. Due to the differences in the spatial location of each LED, the optical signal emitted by each LED experiences different channel conditions. That is, the channel imprints a distinct signature on the signal emitted by a given LED, which makes it unique compared to the signal emitted by other LEDs. Considering an optical MIMO system realised through intensity modulation and direct detection (IM/DD), assuming line-of-sight (LOS) paths

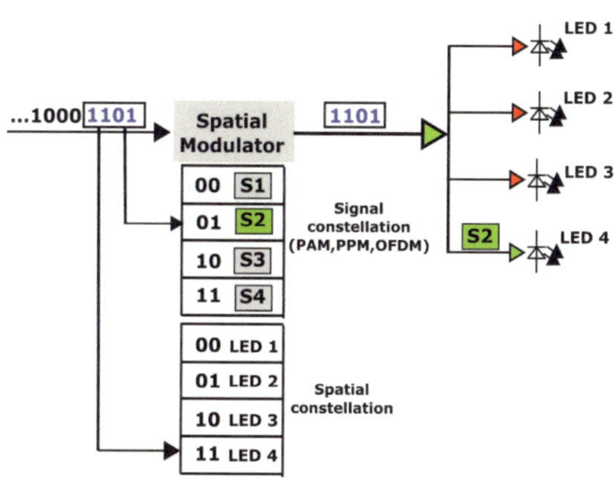

Figure 1. An illustration of the encoding mechanism of an optical SM technique.

between the transmitters and the receivers, with negligible temporal delay [7], the $N_r \times N_t$ optical MIMO channel's gain matrix **H** is given by:

$$\mathrm{H} = \begin{bmatrix} h_{11} & h_{12} & \cdots & h_{1N_t} \\ h_{21} & h_{22} & \cdots & h_{2N_r} \\ \vdots & \vdots & \ddots & \vdots \\ h_{N_r1} & h_{N_r2} & \cdots & h_{N_rN_t} \end{bmatrix} \tag{1}$$

where h_{ij} is the channel's path gain between the i-th PD and the j-th LED. Using the baseband channel model for the VLC system, $(N_r \times 1)$-dimensional vector of the received electrical signal can be expressed as:

$$\mathbf{r}(t) = R\mathbf{H}\mathbf{s}(t) + \mathbf{n}(t), \tag{2}$$

where R is the responsivity of the PD, and $\mathbf{s}(t)$ is the $(N_t \times 1)$-dimensional vector of the transmitted signal. The noise vector $\mathbf{n}(t)$ is the sum of receiver thermal noise and ambient light shot noise; it is usually modelled as independent and identically distributed AWGN.

At the receiver, the detector exploits the uniqueness of the channel condition associated with each transmit-receive channel path to estimate the transmitted data symbol. Thus, a prior knowledge of channel impulse response (CIR) of all the transmit-receive paths is required at the receiver. In practice, the CIR can be obtained through channel estimation technique. According to the maximum likelihood (ML) criterion, the detector computes the Euclidean distance between the received signal and the set of possible signals from all the LN_t combinations of LED index and digital signal modulation index and decides in favour of the combination associated with the smallest Euclidean distance. The estimate of the transmitted signal is obtained as:

$$\hat{\mathbf{S}} = \arg\min_{\mathbf{s} \in \mathbb{S}} \|\mathbf{r} - R\mathbf{H}\mathbf{s}\|^2, \tag{3}$$

where $\mathbb{S}$ is the set of all possible LN_t signals.

2.2. Variants of optical spatial modulation

Due to its promising potentials as highlighted in Section 1, SM has been implemented for VLC systems in various forms. Differences in these variants include, but are not limited to, the inclusion or exclusion of digital signal modulation, the type of digital signal modulation employed, and the number of optical sources that are activated concurrently. Brief descriptions of some reported variants of optical SM are provided as follows.

Optical space shift keying (OSSK): This is a class of optical SM technique, which does not involve the use of digital signal modulation scheme. The source information is encoded solely on the spatial position of the LEDs. As in the conventional optical SM described above, OSSK activates a single LED in a given symbol duration. The total number of bits transmitted per symbol in OSSK is $\log_2(N_t)$. Due to the absence of digital signal modulation, the transceiver requirement such as synchronisation is reduced, and the detection complexity at the receiver is lowered [12]. However, the drawback of not using signal modulation in OSSK is that the achieved transmission rate is lower than conventional SM schemes. This limitation can be addressed by adding signal modulation and/or employing a large array of optical sources and activating multiple sources concurrently. An illustration of OSSK modulation scheme with four LEDs is depicted in **Figure 2**. Two information bits are transmitted per SSK symbol, and the first two bits, '01', are transmitted by activating LED 2.

Generalised spatial shift keying (GSSK): In GSSK, multiple LEDs can be activated during a symbol duration, and information on the transmitted symbol is encoded solely on the indices of the activated LEDs. As in OSSK, signal modulation is not used in GSSK. The GSSK modulation scheme can be implemented by activating a fixed number of sources, N_a, $(1 < N_a < N_t)$ in any symbol duration and varying the indices of the activated sources based on the bits contained in the symbol to be transmitted [13]. The scheme can also be implemented by varying both the number and the indices of the activated sources based on bits of the data symbol to be transmitted [8, 14, 15]. Considering the latter implementation of GSSK, a total of N_t bits are transmitted per data symbol as against $\log_2(N_t)$ in OSSK. As described in Ref. [8], the number and position of ones (1s) in the bits of the symbol to be transmitted determine the number and the indices of the LEDs that will be activated. During a symbol duration, each activated LED transmits a return-to-zero (RZ) pulse of fixed but predefined width and a peak power P_t. For any data symbol with N_t data bits, except when all the bits are zero, the LEDs whose positions correspond to the bit value one are activated to transmit a pulse with duty cycle τ, while all the other LEDs are idle, where $0 < \tau \leq 1$. However, when the data bits are all zero, all the LEDs are activated, but they transmit a pulse with duty cycle $(1 - \tau)$. As an illustration, the pulse pattern for a 2-LED GSSK scheme is shown in **Figure 3**.

Spatial pulse position modulation (SPPM): This is an optical SM scheme that combines SSK with PPM [4]. In SPPM, only one LED is activated to transmit optical data signal in any given symbol duration, and the activated LED transmits an L-PPM pulse pattern, where L is the number of PPM time slots in a symbol duration. By combining SSK with PPM in SPPM, the spectral efficiency of PPM can be improved while still retaining its energy efficiency. In SPPM scheme, a total of $\log_2(LN_t)$ bits are transmitted per symbol. The SPPM scheme is illustrated in **Figure 4** for the case of $N_t = 4$, $L = 2$, and $M = 3$ bits/symbol. The two most significant bits of each symbol determine the index of the LED that will be activated while the last bit determines

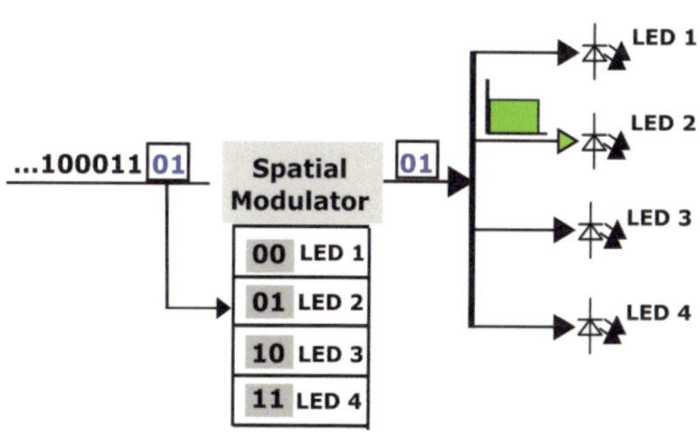

Figure 2. An illustration of OSSK modulation using four LEDs.

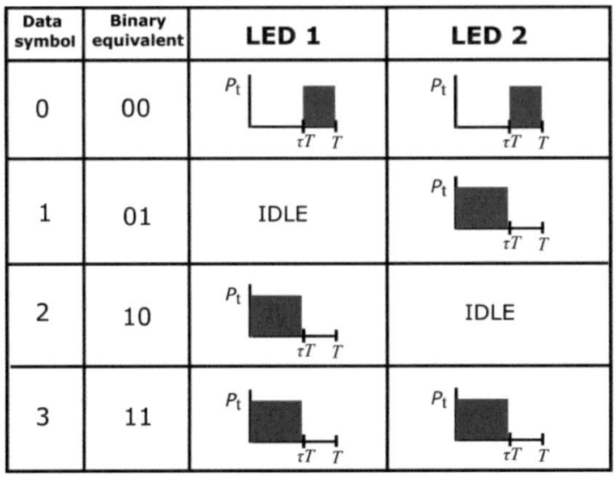

Figure 3. An illustration of the pulse pattern for GSSK scheme using two LEDs.

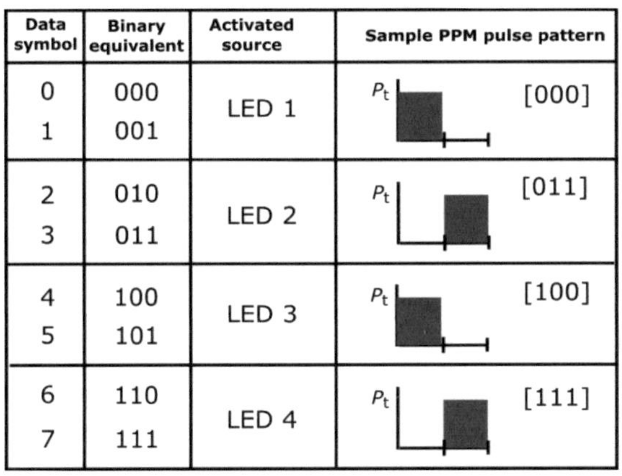

Figure 4. An illustration of intensity modulation for SPPM scheme using four LEDs.

the pulse position in the PPM pulse pattern. As an example, symbol '3' with binary representation '011' is transmitted by activating LED 2 to transmit a pulse in the second time slot.

Spatial pulse amplitude modulation (S-PAM): S-PAM modulation scheme combines SSK and PAM. The encoding mechanism in S-PAM is done in a similar fashion to SPPM. The S-PAM scheme employs a K-PAM scheme for its signal modulation, where K represents the number of optical intensity levels that can be emitted by the activated LED. For instance, the k-th intensity level can be defined as [7]:

$$I_k = \frac{2kI}{K+1}, \text{ for } k = 1, \ldots, K; \tag{4}$$

where I is the mean emitted optical power. A total of $\log_2(LK)$ bits are transmitted during each symbol duration, where the first $\log_2(N_t)$ most significant bits of each symbol determine the index of the LED that will be activated while the remaining $\log_2(K)$ is used to determine the intensity level of the emitted optical radiation.

Generalised spatial modulation (GSM): Like the generalisation of SSK to GSSK, SM schemes can be generalised by activating multiple LEDs in each symbol duration and making each activated LED transmit the same or different digital signal modulation. Such schemes are referred to as generalised spatial modulation (GSM) [5, 16–18]. GSM offers spectral efficiency gain by increasing the number of bits transmitted during each symbol duration. For instance, the GSSK scheme described above can be combined with PPM to develop a generalised SPPM (GSPPM) scheme [5] with a total number of $N_t + \log_2(L)$ bit/symbol. The active sources must, however, be synchronised to avoid inter-symbol interference (ISI).

In the GSPPM scheme [5], during a given symbol duration, one or more LEDs can be activated to concurrently transmit the same L-PPM pulse pattern. The most significant N_t bits of each symbol constitute the spatial constellation point, which determines the indices of the activated LEDs, while the remaining $\log_2(L)$ bits make up the signal constellation point, which is conveyed by the position of the pulse in PPM signal. LED activation in GSPPM is done in a similar fashion to the GSSK scheme, albeit with a slight modification. That is, the number of ones (1s) in the spatial constellation point determines the number and the indices of the active LEDs, but pulse inversion (PI) technique is employed in GSPPM instead of the RZ pulses used in GSSK. The PI technique entails using bipolar pulses such that for all spatial constellation points, except when all the bits that constitute the spatial constellation are zeros, the activated LEDs are driven with electrical pulse signals of amplitude V volts. When all the bits that constitute spatial constellations are zeros, all the LEDs are activated, but they are driven with an electrical pulse signal of amplitude $-V$ volts. By using bipolar pulses, GSPPM requires a DC bias to convert the negative pulses to positive voltages required in IM/DD VLC. Although this implies additional power consumption, however, for applications in VLC, this may not be a drawback since DC bias will be required when the LEDs are used for illumination. The GSPPM scheme is further illustrated in **Figure 5**, for the case of $N_t = 2$ and $L = 2$. As an example, to transmit

Data symbol	Binary equivalent	Spatial Constellation	Signal Constellation	Activated LEDs	Sample PPM Signal Pattern
0	000	00	0	1 and 2	[000]
1	001	00	1		
2	010	01	0	1	[011]
3	011	01	1		
4	100	10	0	2	[100]
5	101	10	1		
6	110	11	0	1 and 2	[111]
7	111	11	1		

Figure 5. An illustration of GSPPM scheme.

symbol '1', with binary equivalent '001', because the first two bits which are used to select the spatial constellation point are both zeros, the two LEDs are activated and are driven by $-V$ volts while the last bit of the symbol, '1', indicates that the pulse be transmitted in the second time slot.

2.3. Contrast with spatial multiplexing, spatial diversity, and repetition coding

Beside SM, other MIMO transmission techniques that have been considered for VLC include repetitive coding (RC) and spatial multiplexing (SMP). In RC, the same information is simultaneously transmitted from all the transmitters, and the transmitted signals add up constructively at the receiver. In essence, RC offers diversity gain, which makes it more robust to channel the correlation compared to SMP and SM. However, since RC does not provide spatial multiplexing gains, large signal constellation sizes will be required to achieve high spectral efficiency. In contrast, SMP enables high data rates by transmitting different kinds of information from each transmitter. The drawback of SMP is that it requires sufficiently low channel correlation. SM is more robust to correlated channels compared to SMP, and it provides larger spectral efficiency compared to RC [7].

In terms of complexity, the optical SM constitutes a low complexity technique of increasing the achievable transmission rates in VLC systems. Compared to other spectrally efficient modulation schemes like the optical OFDM [19] and PAM, the SM technique is not as sensitive to nonlinearity effects of the system components [4, 20]. Hence, SM-based VLC systems do not require the complex pre-distortion algorithm to compensate for device nonlinearity. Moreover, since only one or a few LED(s) is active in each symbol duration, inter-channel interference (ICI), which results from multiple and concurrent signal transmission, is reduced in optical SM as compared to RC and SMP. Hence, receiver design is made simpler and ML-optimum performance can be achieved at a reduced decoding complexity [21].

The computational complexity of the ML-based detection of optical S-PAM is compared with that of RC and SMP in **Table 1** [22]. The computational complexity is defined as the total number of required mathematical operations, that is, multiplications, additions, and subtractions that are required for ML detection. **Table 1** shows that, to achieve equal spectral efficiency, the detection of SM is less computationally intensive compared to RC and SMP. Because SM conveys additional bits via the spatial domain, it employs a smaller digital signal constellation size to achieve the same spectral efficiency as RC.

MIMO technique	Number of mathematical operations at the receiver
RC	$L(2N_tN_r + N_r - 1)$
SMP	$L^{N_t}(2N_tN_r + N_r - 1)$
SM	$LN_t(3N_r - 1)$

L denotes the size of digital signal constellation.

Table 1. Computational complexity at the receiver of RC, SMP, and SM techniques [22].

3. Error performance of optical spatial modulation

In this section, the error performance of four variants of optical SM technique is analysed and closed-form expressions for their symbol error are derived. These expressions are validated using Monte-Carlo simulations. The optical SM variants considered are SPPM, OSSK, GSSK, and GSPPM. In the following, without any loss of generality, a MIMO VLC system, $N_t = 4$ and $N_r = 2$, is considered. The LOS channel gain matrix, **H**, is obtained from the simulated indoor optical channel, which is performed based on the ray-tracing algorithm in Ref. [23]. The normalised LOS channel matrix is obtained as:

$$\mathbf{H} = \begin{bmatrix} 1 & 0.3118 & 0.2496 & 0.1283 \\ 1 & 0.3925 & 0.2709 & 0.1578 \end{bmatrix} \tag{5}$$

3.1. Error performance analysis of SPPM

As described in Section 2.2, the data symbol in SPPM is conveyed by the index of the active LED and the pulse position of the transmitted PPM signal. These two parameters must be estimated at the receiver in order to demodulate the transmitted symbol. Hence, error performance analysis will involve evaluating the probability of the correct pulse position and LED index detection. Considering that an SPPM symbol is transmitted by activating the j-th LED to transmit a pulse in the m-th PPM time slot, the transmit signal vector $\mathbf{s}_j^m(t)$ can be expressed as:

$$\mathbf{s}_j^m(t) = [0, \cdots, x_j^m(t), \cdots, 0]^{\mathrm{T}}. \tag{6}$$

The nonzero entry, $x_j^m(\mathrm{t})$, positioned at the index of the active j-th LED, is the L-PPM signal imposed on the optical carrier emitted by the active LED. Without loss of generality, using rectangular pulse shaping, the L-PPM signal with amplitude P_t can be defined by:

$$x_j^m(t) = \begin{cases} P_t; & \text{for}(m-1)T_c \leq t \leq mT_c \\ 0\,; & \text{elsewhere} \end{cases} \tag{7}$$

where $T_c = T/L$ is the duration of each PPM time slot and T is the symbol duration. The receivers employ a unit energy receive filter, $f_c(t)$, which is matched to the PPM waveform and is given by:

$$f_c(t) = \frac{1}{\sqrt{T_c}} \mathrm{rect}\left(\frac{t - T_c/2}{T_c}\right). \tag{8}$$

where $\mathrm{rect}(x) = 1$ for $|x| < 1/2$. The outputs of the matched filters (MF) are sampled at the rate $1/T_c$ to obtain the samples of each PPM time slot. For N_r PDs and L-PPM, the $N_r \times L$ array of the outputs of the matched filters in each time slot and for all PDs can be expressed as:

$$\begin{aligned} \mathbf{Y} &= \mathbf{X}_j^m + \mathbf{N} \\ &= [\mathbf{0} \quad \cdots \mathbf{h}_j \sqrt{E_S^{\mathrm{SPPM}}} \cdots \mathbf{0}] + [\mathbf{n}_1 \quad \cdots \mathbf{n}_m \cdots \mathbf{n}_L] \\ &= [\mathbf{y}_1, \quad \cdots \mathbf{y}_m, \ \cdots \mathbf{y}_L] \end{aligned} \tag{9}$$

where $\mathbf{X}_j^m$ is the $(N_r \times L)$-dimensional array of the samples of noiseless matched filter output and $\mathbf{N}$ is the Gaussian noise at the output of the matched filter with variance $\sigma_n^2 = N_0/2$. $E_s^{SPPM} = (RP_t)^2 T_c$ is the transmit energy of the SPPM symbol. The column vector $\mathbf{h}_j$ represents the channel gain from the j-th LED to all the PDs, that is, the j-th column of the channel matrix $\mathbf{H}$. Using ML detection criterion, the estimate of transmitted symbol, $\hat{a} = [\hat{m}, \hat{j}]$, is obtained as:

$$\hat{a} = [\hat{m}, \hat{j}] = \arg\max_{m,j} p(\mathbf{Y}|\, \mathbf{X}_j^m) \tag{10}$$

$$[\hat{m}, \hat{j}] = \arg\max_{m,j} D(\mathbf{Y}|\, \mathbf{X}_j^m) \tag{11}$$

The joint probability density function of $\mathbf{Y}$ conditioned on $\mathbf{X}_j^m$ is given by:

$$p(\mathbf{Y}|\, \mathbf{X}_j^m) = \frac{1}{(\pi N_0)^{LN_r/2}} \exp\left[-\frac{\|\mathbf{Y} - \mathbf{X}_j^m\|_F^2}{N_0}\right], \tag{12}$$

and the Euclidean distance metric $\mathrm{D}(\mathbf{Y}, \mathbf{X}_j^m)$ is computed as:

$$\mathrm{D}(\mathbf{Y}, \mathbf{X}_j^m) = \|\mathbf{Y} - \mathbf{X}_j^m\|_F^2 \tag{13}$$

where $\|\cdot\|_F$ denotes the Frobenius norm. The detection process entails obtaining the estimated pulse position, $\widehat{m}$, from:

$$\hat{m} = \arg\max_m \left(\sum_i^{N_r} y_{i,\,m}\right), \tag{14}$$

and then estimating the index of the activated LED from the minimum Euclidean distance metric using the ML detection criterion.

Let the probability of a correctly decoded pulse position be defined by $P_{c,\,ppm} = p(\hat{m} = m)$; then, the probability of correctly decoding the LED index given that the pulse position has been correctly decoded can be expressed as $P_{c,\,tx} = p(\hat{j} = j \mid \hat{m} = m)$. Therefore, the probability of correct symbol detection is given by:

$$P_{c,\,sym}^{SPPM} = P_{c,\,tx} \times P_{c,\,ppm} = p(\hat{j} = j \mid \hat{m} = m) \times p(\hat{m} = m) \tag{15}$$

and the probability of symbol error is obtained as:

$$P_{e,\,sym}^{SPPM} = 1 - P_{c,\,sym}^{SPPM} = 1 - (P_{c,\,tx} \times P_{c,\,ppm}) \tag{16}$$

Probability of correctly detecting the index of the activated LED. To evaluate the probability of correctly decoding the LED index, we shall first find the pairwise error probabilities (PEP). Consider that the activated LED transmitted a pulse in slot m, the $\mathrm{PEP}_m^{j\to k}$, that the receiver decides in favour of LED k instead of LED j is given by:

$$\mathrm{PEP}_m^{j\to k} = p\left(D(\mathbf{Y},\mathbf{X}_j^m) > D(\mathbf{Y},\mathbf{X}_k^m)\right) = p\left(\left\|\mathbf{y}_m - \mathbf{h}_j\sqrt{E_\mathrm{S}^\mathrm{SPPM}}\right\|_\mathrm{F}^2 > \left\|\mathbf{y}_m - \mathbf{h}_k\sqrt{E_\mathrm{S}^\mathrm{SPPM}}\right\|_\mathrm{F}^2\right)$$

$$= \mathrm{Q}\left(\|\mathbf{h}_j - \mathbf{h}_k\|_F\sqrt{\frac{E_\mathrm{S}^\mathrm{SPPM}}{2N_0}}\right) = \mathrm{Q}\left(\|\mathbf{h}_j - \mathbf{h}_k\|_F\sqrt{\frac{\gamma_\mathrm{S}^\mathrm{SPPM}}{2}}\right) \tag{17}$$

where $\mathrm{Q}(\cdot)$ denotes the Q-function, and $\gamma_\mathrm{S}^\mathrm{SPPM} = E_\mathrm{S}^\mathrm{SPPM}/N_0$ is the transmit signal-to-noise ratio (SNR) per SPPM symbol. For N_t equiprobable LEDs, the probability of error in decoding the index of the activated LED conditioned on a correctly decoded pulse position is obtained as:

$$P_\mathrm{e,\,tx} = \frac{2}{N_\mathrm{t}}\sum_{j=1}^{N_\mathrm{t}-1}\sum_{k=j+1}^{N_\mathrm{t}} \mathrm{PEP}_m^{j\to k} = \frac{2}{N_\mathrm{t}}\sum_{j=1}^{N_\mathrm{t}-1}\sum_{k=j+1}^{N_\mathrm{t}} \mathrm{Q}\left(\|\mathbf{h}_j - \mathbf{h}_k\|_\mathrm{F}\sqrt{\frac{\gamma_\mathrm{S}^\mathrm{SPPM}}{2}}\right), \tag{18}$$

and the probability of correctly decoding the index of the activated LED is computed as:

$$P_\mathrm{c,\,tx} = 1 - \left[\frac{2}{N_t}\sum_{j=1}^{N_t-1}\sum_{k=j+1}^{N_t} \mathrm{Q}\left(\|\mathbf{h}_j - \mathbf{h}_k\|_\mathrm{F}\sqrt{\frac{\gamma_s^\mathrm{SPPM}}{2}}\right)\right] \tag{19}$$

Probability of correctly detecting the transmitted pulse position. Given that the j-th LED is activated to transmit a pulse in slot m, then the $\mathrm{PEP}^j_{m\to q}$ that the receiver decides in favour of slot q instead of slot m can be expressed as:

$$\mathrm{PEP}^j_{m\to q} = p\left(\mathfrak{D}(\mathbf{Y},\mathbf{X}_j^m) > D(\mathbf{Y},\mathbf{X}_j^q)\right) \tag{20}$$

where,

$$\mathfrak{D}(\mathbf{Y},\mathbf{X}_j^m) = \left\|\mathbf{y}_\mathrm{m} - \mathbf{h}_j\sqrt{E_\mathrm{S}^\mathrm{SPPM}}\right\|_\mathrm{F}^2 + (\mathbf{y}_q)^2 = (\mathbf{n}_m)^2 + (\mathbf{n}_q)^2 \tag{21}$$

$$\mathfrak{D}(\mathbf{Y},\mathbf{X}_j^q) = (\mathbf{y}_m)^2 + \left\|\mathbf{y}_q - \mathbf{h}_j\sqrt{E_\mathrm{S}^\mathrm{SPPM}}\right\|_\mathrm{F}^2 = \left\|\mathbf{h}_j\sqrt{E_\mathrm{S}^\mathrm{SPPM}} + \mathbf{n}_m\right\|_\mathrm{F}^2 + \left\|\mathbf{n}_q - \mathbf{h}_j\sqrt{E_\mathrm{S}^\mathrm{SPPM}}\right\|_\mathrm{F}^2 \tag{22}$$

Therefore,

$$\mathrm{PEP}^j_{m\to q} = p\left((\mathbf{n}_q - \mathbf{n}_m)^\mathrm{T}\mathbf{h}_j > \sqrt{E_\mathrm{S}^\mathrm{SPPM}}\|\,\mathbf{h}_j\,\|_\mathrm{F}^2\right) = \mathrm{Q}(\|\mathbf{h}_j\|_\mathrm{F}\sqrt{\gamma_\mathrm{S}^\mathrm{SPPM}}). \tag{23}$$

For N_t equiprobable LEDs, the union bound technique [24] gives the probability of error in decoding the transmitted pulse position as:

$$P_{e,\,ppm} = \frac{1}{N_t}\sum_{j=1}^{N_t}\left((L-1)\times \mathrm{PEP}^{j}_{m\to q}\right) = \frac{1}{N_t}\sum_{j=1}^{N_t}\left((L-1)\times \mathrm{Q}(\|\mathbf{h}_j\|_F\sqrt{\gamma_S^{SPPM}})\right) \tag{24}$$

Thus, the average probability of correctly decoding the transmitted pulse position is:

$$P_{c,\,ppm} = 1 - \left[\frac{1}{N_t}\sum_{j=1}^{N_t}\left((L-1)\times \mathrm{Q}\left(\|\mathbf{h}_j\|_F\sqrt{\gamma_S^{SPPM}}\right)\right)\right] \tag{25}$$

By combining Eqs. (16), (19), and (25), the union bound on the average probability of symbol error of the SPPM scheme is derived as [4]:

$$P_{e,\,sym}^{SPPM} \le 1 - \left[1 - \frac{2}{N_t}\sum_{j=1}^{N_t-1}\sum_{k=j+1}^{N_t}\mathrm{Q}\left(\|\mathbf{h}_j - \mathbf{h}_k\|_F\sqrt{\frac{\gamma_S^{SPPM}}{2}}\right)\right]$$
$$\left[\frac{(L-1)}{N_t}\sum_{j=1}^{N_t}\left(\frac{1}{(L-1)} - \mathrm{Q}(\|\mathbf{h}_j\|_F\sqrt{\gamma_S^{SPPM}})\right)\right] \tag{26}$$

The expression in Eq. (26) indicates that for a given SNR, the error performance of SPPM strongly depends on the individual channel path gain of each transmitter-receiver link as well as the difference between these channel gain values. The error performance plots for different system configurations are shown in **Figure 6**. For the case of $N_r = 1$, $N_t = [2, 4]$, and $L = [2, 8]$, the achieved symbol error rate (SER) is plotted against the SNR per bit $\gamma_b = \gamma_s/M$. To validate the analysis, simulation results are also included. **Figure 6** shows that the analytical bound in Eq. (26) agrees with the simulation results. The slight deviation observed at SER > 10^{-2} is due to the union bound techniques used in the analysis.

Moreover, as L increased from 2 to 8, the γ_b required to achieve an SER of 10^{-6} reduced by about 2 dB for $N_t = [2, 4]$. This shows the energy efficiency benefit that PPM adds to SPPM. Using more LEDs results in higher attainable data rate, but this requires distinct channel gains for all the LEDs and higher SNR. This explains why using two LEDs require up to about 16 dB less in SNR than using four LEDs, as shown in **Figure 6**.

3.2. Error performance analysis of OSSK

Using the description of the OSSK scheme provided in Section 2.2, OSSK can be viewed as a subset of SPPM scheme in which $L = 1$, and $T_c = T$. The transmitted symbol in SSK is determined by simply estimating the index of the activated LED. Therefore, the error performance analysis of OSSK is equivalent to the process of evaluating the probability of error in detecting the active LED in SPPM. By using $L = 1$ in the formulation of the error performance analysis of SPPM, the union bound on the average probability of symbol error of the SSK modulation scheme is obtained as:

$$P_{\text{e, sym}}^{\text{SSK}} \leq \frac{2}{N_t} \sum_{j=1}^{N_t-1} \sum_{k=j+1}^{N_t} Q\left(\|\mathbf{h}_j - \mathbf{h}_k\|_F \sqrt{\frac{\gamma_S^{\text{SSK}}}{2}} \right) \tag{27}$$

where $\gamma_s^{\text{SSK}} = E_s^{\text{SSK}}/N_0$ is the transmit SNR per SSK symbol. $E_s^{\text{SSK}} = (RP_t)^2T$ is the transmit energy per symbol. Note that Eq. (27) is the same as Eq. (18), and it can also be obtained by simply substituting $L = 1$ in Eq. (26). This conforms with our description of OSSK as being equivalent to an SPPM scheme with $L = 1$.

Using the closed-form expression in Eq. (27), the SER of OSSK is plotted against the SNR per bit γ_b in **Figure 7**. The plot shows a very tight match between the simulation and the theoretical analysis. It can be observed that at an SER of 10^{-6}, compared to $N_r = 1$, using $N_r = 2$ provides an SNR gain of about 2.5 and 7 dB for $N_t = 2$ and $N_r = 4$, respectively. This implies that the dependency of optical SM on decorrelated channel can be relaxed by using multiple PDs.

3.3. Error performance analysis of GSSK

According to the GSSK scheme described in Section 2.2, during a symbol duration, the number and the indices of the active LEDs are determined by the bits of the data symbol to be transmitted. Using N_t LEDs, there are 2^{N_t} possible GSSK symbols. Considering a single symbol duration, T, in which symbol v, for $v = 0, \ldots, (2^{N_t} - 1)$, is transmitted, the $(N_t \times 1)$-dimensional vector of the transmitted signal, $\mathbf{s}(t)$, can be expressed as $\mathbf{s}(t) = \tau P_t \mathbf{a}_v$, where $\mathbf{a}_v = [a_1, \ldots, a_{N_t}]^T$ is the LED activation vector for symbol v, and τ is the duty cycle of the transmitted optical pulse as explained in Section 2.2. The entries of $\mathbf{a}_v$ are binary digits, with

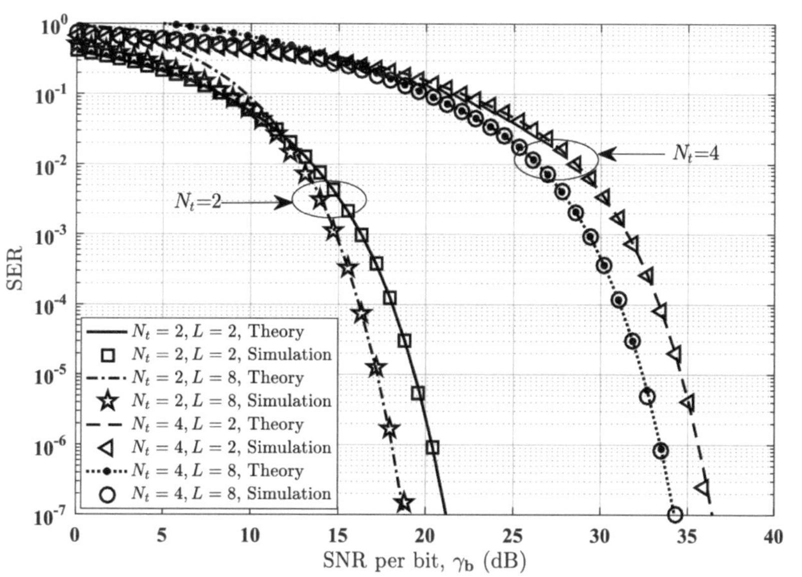

Figure 6. Error performance plot of SPPM for $N_t = [2,4]$, $N_r = 1$, and $L = [2, 8]$.

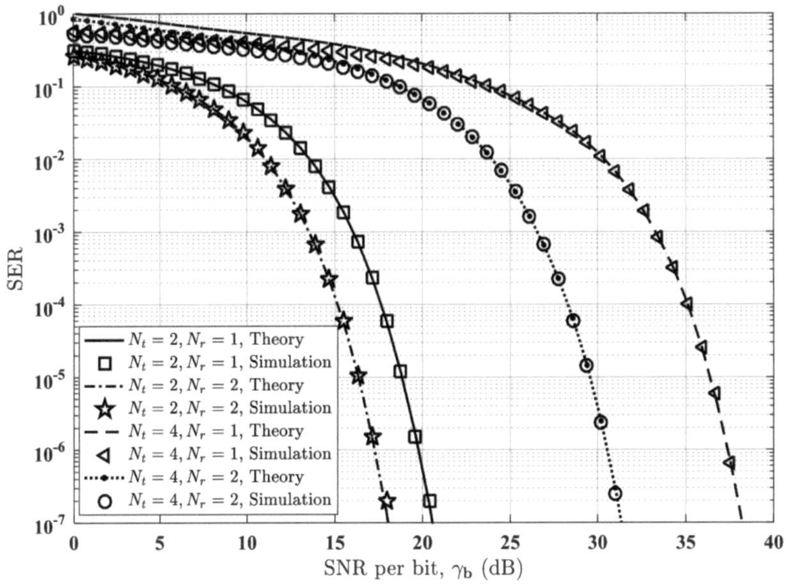

Figure 7. Error performance plot of OSSK for N_t = [2, 4] and N_r = [1, 2].

'1s' as the indices of activated LEDs and '0s' as the indices of the idle LEDs. The $(N_r \times 1)$-dimensional vector of the received electrical signal, $\mathbf{r}(t)$, is given by:

$$\mathbf{r}(t) = \tau R P_t \mathbf{H}\mathbf{a}_v + \mathbf{n}(t), \text{ for } 0 < t < T. \tag{28}$$

For tractability, $\mathbf{w}_v \triangleq \mathbf{H}\mathbf{a}_v$ is defined as the vector of the channel gains associated with the transmission of the GSSK symbol v. Hence,

$$\mathbf{r}(\mathrm{t}) = \tau R P_t \mathbf{w}_\mathrm{v} + \mathbf{n}(\mathrm{t}), \text{ for } 0 < t < T. \tag{29}$$

As an illustration, considering a 2-LED GSSK system with N_r PDs, the received signal for all possible GSSK symbols is obtained as shown in **Table 2**. To estimate the transmitted symbol, the output of the matched filter, sampled at the rate $1/T$, is given by:

$$\mathbf{y} = \mathbf{x}_v + \mathbf{n} \tag{30}$$

where $\mathbf{x}_v = \mathbf{w}_v \sqrt{E_S^{GSSK}}$ and energy per symbol $E_S^{GSSK} = (RP_t)^2 \tau T$. Considering that GSSK symbol v is transmitted, the pairwise error probabilities, $\mathrm{PEP}^{v \to u}$, that the receiver decides in favour of symbol u instead of symbol v are expressed as:

v	**Binary equivalent**	**Activated LED**	w_v	**r(*t*)**
0	00	LED 1 and LED 2	$\mathbf{h}_1 + \mathbf{h}_2$	$(1-\tau)RP_t(\mathbf{h}_1 + \mathbf{h}_2) + \mathbf{n}(t)$
1	01	LED 1	$\mathbf{h}_1$	$\tau RP_t\mathbf{h}_1 + \mathbf{n}(t)$
2	10	LED 2	$\mathbf{h}_2$	$\tau RP_t\mathbf{h}_2 + \mathbf{n}(t)$
3	11	LED 1 and LED 2	$\mathbf{h}_1 + \mathbf{h}_2$	$\tau RP_t(\mathbf{h}_1 + \mathbf{h}_2) + \mathbf{n}(t)$

Table 2. LED activation for GSSK modulation using two LEDs.

$$\begin{aligned}\mathrm{PEP}^{v\rightarrow u} &= p\Big(D(\mathbf{y},\mathbf{x}_v) > D(\mathbf{y},\mathbf{x}_u)\Big)\\ &= p\left(\left\|\mathbf{y}-\mathbf{w}_v\sqrt{E_\mathrm{S}^\mathrm{GSSK}}\right\|_\mathrm{F}^2 > \left\|\mathbf{y}-\mathbf{w}_u\sqrt{E_\mathrm{S}^\mathrm{GSSK}}\right\|_\mathrm{F}^2\right)\\ &= \mathrm{Q}\left(\|\mathbf{w}_v-\mathbf{w}_u\|_\mathrm{F}\sqrt{\frac{E_\mathrm{S}^\mathrm{GSSK}}{2N_0}}\right) = \mathrm{Q}\left(\|\mathbf{w}_v-\mathbf{w}_u\|_\mathrm{F}\sqrt{\frac{\gamma_\mathrm{S}^\mathrm{GSSK}}{2}}\right).\end{aligned} \tag{31}$$

where $\gamma_\mathrm{s}^\mathrm{GSSK} = E_\mathrm{s}^\mathrm{GSSK}/N_0$ is the transmit SNR per GSSK symbol. For 2^{N_t} equiprobable GSSK symbols, the union bound on the probability of symbol error of GSSK modulation is given by [8]:

$$P_\mathrm{e,\,sym}^\mathrm{GSSK} \le \frac{2}{2^{N_\mathrm{t}}}\sum_{v=1}^{2^{N_\mathrm{t}}-1}\sum_{u=v+1}^{2^{N_\mathrm{t}}}\mathrm{PEP}^{v\rightarrow u} = \frac{2}{2^{N_\mathrm{t}}}\sum_{v=1}^{2^{N_\mathrm{t}}-1}\sum_{u=v+1}^{2^{N_\mathrm{t}}}\mathrm{Q}\left(\|\mathbf{w}_v-\mathbf{w}_u\|_\mathrm{F}\sqrt{\frac{\gamma_\mathrm{S}^\mathrm{GSSK}}{2}}\right). \tag{32}$$

Without loss of generality, using $\tau = 1$, the closed-form expression derived for the SER of GSSK in Eq. (32) is validated by tightly matched simulation results as depicted by the error performance plots in **Figure 8** for the cases N_t = [2,4] and N_r = [1,2].

3.4. Error performance analysis of GSPPM

The performance analysis of GSPPM scheme entails combining the detection process in GSSK with the pulse position detection in SPPM. The transmitted symbol consists of the spatial constellation point, which determines the active LED and the pulse position of the transmitted PPM signal. Let λ denote pulse inversion constant associated with each spatial constellation point; $\lambda = -1$ if all the bits that constitute the spatial constellation point of the data symbol are zeros and otherwise $\lambda = +1$. Considering a single symbol duration, if the active LEDs transmit a pulse in the m-th time slot, the $(N_\mathrm{r} \times 1)$-dimensional vector of the received electrical signal is given by:

$$\mathbf{r}(t) = \Big(R\lambda_v x_j^m(t)\Big)\mathbf{w}_v + \mathbf{n}(t),\ \text{for } 0 < t < T. \tag{33}$$

where v denotes the spatial constellation point of the transmitted symbol. Other variables are as defined previously. As an illustration, considering a two-LED GSPPM system, the signal received in the pulse position of the transmitted PPM signal is obtained as shown in **Table 3**.

The error performance analysis will involve evaluating the probability of correctly estimating the spatial constellation point and the pulse position. The detection of the transmitted spatial constellation in GSPPM is equivalent to the detection of the transmitted GSSK symbol in Section 3.3. Hence, the probability of correctly detecting the transmitted spatial constellation is obtained as [5]:

$$P_\mathrm{c,\,scp} \le 1 - \frac{2}{2^{N_\mathrm{t}}}\sum_{v=1}^{2^{N_\mathrm{t}}-1}\sum_{u=v+1}^{2^{N_\mathrm{t}}}\mathrm{Q}\left(\|\lambda_v\mathbf{w}_v-\lambda_u\mathbf{w}_u\|_\mathrm{F}\sqrt{\frac{\gamma_\mathrm{S}^\mathrm{GSPPM}}{2}}\right) \tag{34}$$

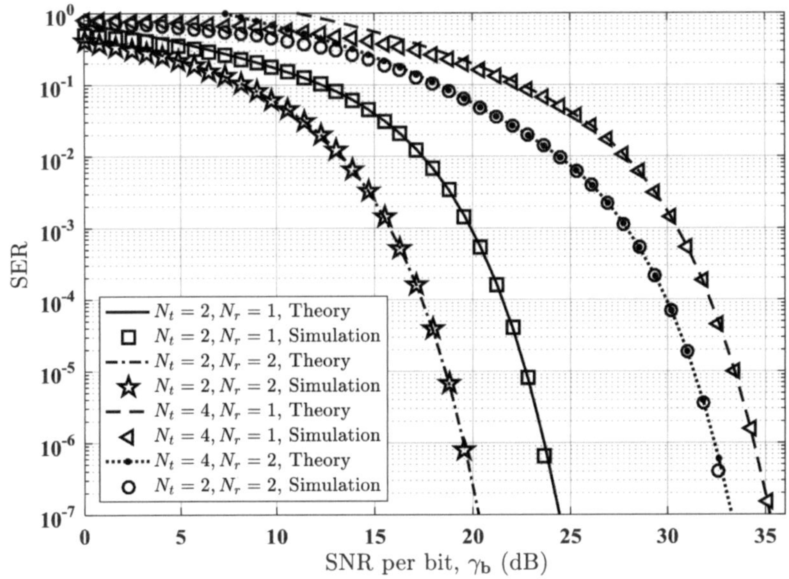

Figure 8. Error performance plot of GSSK for N_t = [2, 4] and N_r = [1, 2].

Spatial constellation	Activated LED	$\mathbf{w}_v$	λ_v	r(t)
00	LED 1 and LED 2	$\mathbf{h}_1 + \mathbf{h}_2$	−1	$-RP_t(\mathbf{h}_1 + \mathbf{h}_2) + \mathbf{n}(t)$
01	LED 1	$\mathbf{h}_1$	1	$RP_t\mathbf{h}_1 + \mathbf{n}(t)$
10	LED 2	$\mathbf{h}_2$	1	$RP_t\mathbf{h}_2 + \mathbf{n}(t)$
11	LED 1 and LED 2	$\mathbf{h}_1 + \mathbf{h}_2$	1	$RP_t(\mathbf{h}_1 + \mathbf{h}_2) + \mathbf{n}(t)$

Table 3. LED activation and received signal for two-LED GSSPM system.

where $\gamma_{\rm S}^{\rm GSPPM} = E_{\rm S}^{\rm GSPPM}/N_0$ is the transmit SNR per GSPPM symbol and $E_{\rm S}^{\rm GSPPM} = (RP_{\rm t})^2 T_c$ transmit energy of GSSPM symbol. Moreover, using Eq. (25), the average probability of correctly decoding the transmitted pulse position can be expressed as [5]:

$$P_{\rm c,\,ppm}^{\rm GSPPM} = 1 - \left[\frac{1}{2^{N_t}}\sum_{v=1}^{2^{N_t}}\left((L-1)\times Q\left(\|\lambda_v \mathbf{w}_v\|_F\sqrt{\gamma_{\rm S}^{\rm GSPPM}}\right)\right)\right] \tag{35}$$

By combining Eqs. (34) and (35), the union bound on the average probability of symbol error of the GSPPM scheme is derived as [5]:

$$\begin{aligned} P_{\rm e,\,sym}^{\rm GSPPM} \leq 1 - &\left[1 - \frac{2}{2^{N_{\rm t}}}\sum_{v=1}^{2^{N_{\rm t}}-1}\sum_{u=v+1}^{2^{N_{\rm t}}} Q\left(\|\lambda_v \mathbf{w}_v - \lambda_u \mathbf{w}_u\|_F\sqrt{\frac{\gamma_{\rm S}^{\rm GSPPM}}{2}}\right)\right] \\ &\times \left[\frac{(L-1)}{2^{N_{\rm t}}}\sum_{v=1}^{2^{N_{\rm t}}}\left(\frac{1}{(L-1)} - Q\left(\|\lambda_v \mathbf{w}_v\|_F\sqrt{\gamma_{\rm S}^{\rm GSPPM}}\right)\right)\right] \end{aligned} \tag{36}$$

As shown in **Figure 9**, the closed-form expression for the theoretical SER of GSPPM in Eq. (36) is validated by simulation results. Moreover, the plot also highlights the energy efficiency benefit that PPM adds to GSPPM. As L increases from 2 to 8, the γ_b required to achieve an SER of 10^{-6} reduces by about 2 and 1.5 dB for $N_t = 2$ and $N_t = 4$, respectively.

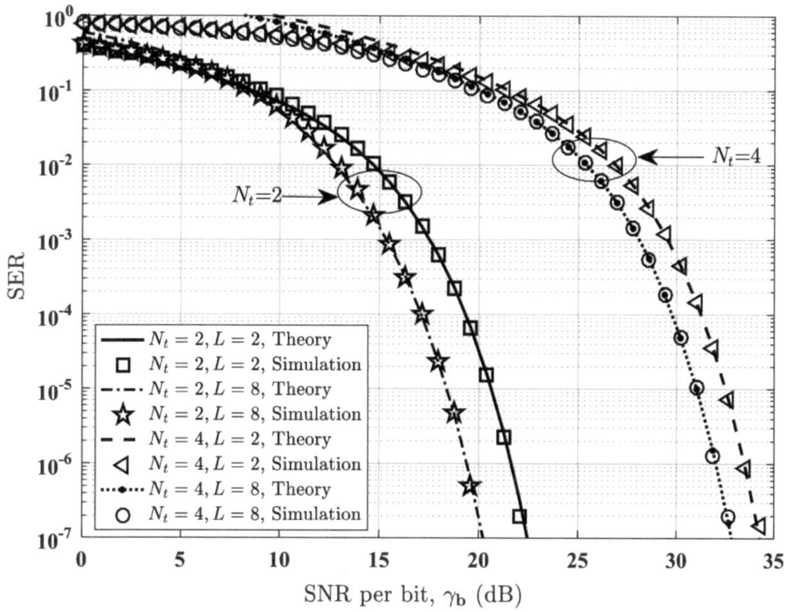

Figure 9. Error performance plot of GSPPM for N_t = [2, 4], N_r = 1, and L = [2, 8].

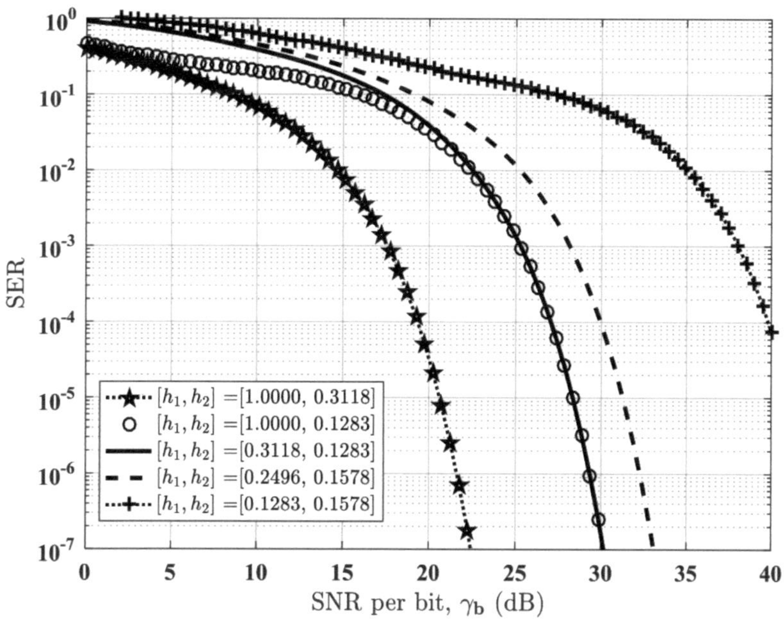

Figure 10. Error performance plot of GSPPM for different channel gains, N_t = 2 and L = 2.

Using GSPPM as a case study, the impact of channel gains on the performance of optical SM is illustrated in **Figure 10** with the error performance plot for different channel gain values. The plots show that the higher the difference between the channel gains, the lower the SNR required to achieve a given SER. According to Eq. (36), the SER also depends on the absolute value of the channel gains. For instance, to achieve an SER of 10^{-6}, the normalised channel gain $[h_1, h_2]$ = [1, 0.3118] requires about 20 dB less in SNR than the set $[h_1, h_2]$ = [1, 0.1283] even though the channel gain separation $|h_1, h_2|$ for the latter set is higher. Moreover, the effect of the individual channel gain values becomes more significant as SNR increases. This is observed in the case of the normalised channel gain sets $[h_1, h_2]$ = [1, 0.1283] and $[h_1, h_2]$ = [0.3118, 0.283]. Both sets require about 29 dB to achieve an SER of 10^{-6}. It can thus be inferred from **Figure 10** that at lower SNR values, the error performance is dictated by the difference in the channel gains and at high SNR by the value of the smaller channel gain.

4. SM comparison with conventional modulation techniques

A performance comparison among different variants of optical SM and other common modulation schemes used for optical wireless communications is shown in **Table 4**. These modulation schemes are compared based on the average transmitted optical power P_{ave}, spectral efficiency η_{spec} and the number of bits per symbol M. These parameters can be obtained from the expressions provided in **Table 4**, where $P_{\text{t}}^{\text{modtype}}$ represents the transmitted peak optical power for the stated modulation type 'modtype', and τ is the duty cycle of the modulation schemes that utilise RZ pulse pattern. The spectral efficiency here is defined as the ratio of the bit rate to the bandwidth requirement of the system, where bandwidth requirement is equivalent to the reciprocal of the pulse duration in each of the modulation scheme.

Considering the case of N_t = 2 and using the expressions in **Table 4**, the spectral and energy efficiency of the four optical SM schemes analysed in Section 3 are plotted against the number of PPM time slots, L, in **Figures 11** and **12**, respectively. Energy efficiency is expressed in terms of the SNR required to achieve a representative SER of 10^{-6}, and for a fair comparison, all the modulation techniques are assumed to have the same P_{ave} as the GSPPM. For the configuration

Modulation type	P_{ave}	M	η_{spec}
RC RZ-OOK	$\frac{\tau}{2}P_{\text{t}}^{\text{OOK}}$	1	τ
RC L-PPM	$\frac{1}{L}P_{\text{t}}^{\text{PPM}}$	$\log_2 L$	$\frac{1}{L}\log_2 L$
SSK	$P_{\text{t}}^{\text{SSK}}$	$\log_2 \text{N}_{\text{t}}$	$\log_2 N_{\text{t}}$
GSSK	$N_{\text{t}}P_{\text{t}}^{\text{GSSK}}\left(\frac{\tau}{2}+\frac{1-\tau}{2^{N_{\text{t}}}}\right)$	N_{t}	N_{t}
SPPM	$\frac{P_{\text{t}}^{\text{SPPM}}}{L}$	$\log_2(LN_{\text{t}})$	$\frac{\log_2(LN_{\text{t}})}{L}$
GSPPM	$\frac{N_{\text{t}}P_{\text{t}}^{\text{GSPPM}}}{L}\left(\frac{1}{2}+\frac{1}{2^{N_{\text{t}}}}\right)$	$\text{N}_{\text{t}}+\log_2 L$	$\frac{\text{N}_{\text{t}}+\log_2 L}{L}$

Table 4. Comparison of optical modulation schemes.

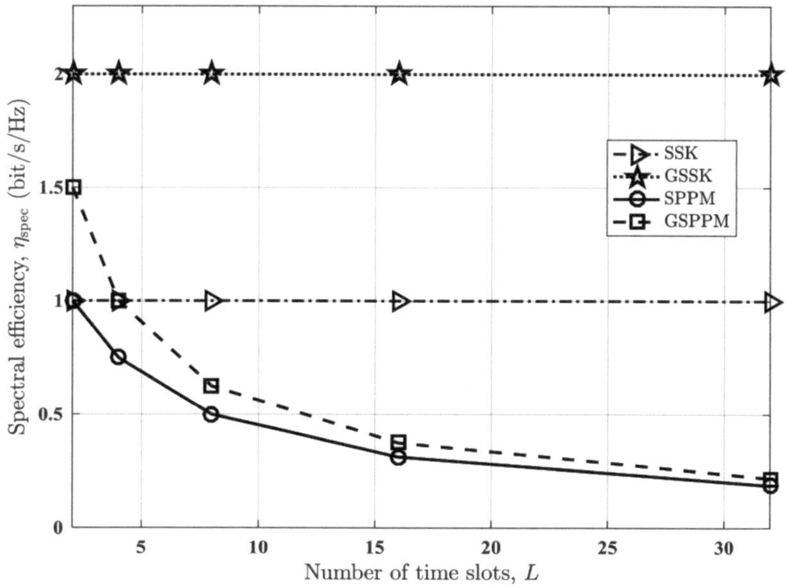

Figure 11. Spectral efficiency comparison of optical SM schemes.

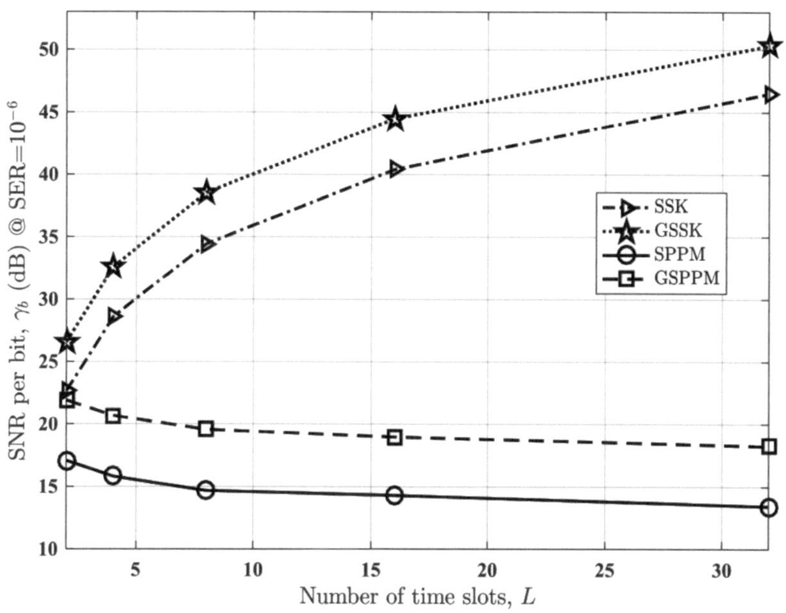

Figure 12. Energy efficiency comparison of optical SM schemes.

described above, as shown in **Figure 11**, the η_{spec} of GSSK exceeds the maximum achievable values in SSK and SPPM by 1 bits/s/Hz and that of the GSPPM scheme by 0.5 bits/s/Hz. The superiority of GSSK over GSPPM in terms of η_{spec} is because the pulse duration in GSSK is L times longer than that of GSPPM, even though GSPPM transmits more bits/symbol. However, in terms of energy efficiency, **Figure 12** shows that GSSK and SSK are less energy efficient when compared with SPPM and GSPPM. The SNR requirement of GSSK for $L = 8$ exceeds those of SPPM and GSPPM by up to 23 and 18 dB, respectively. This can be attributed to the energy efficiency benefit that PPM adds to both SPPM and GSPPM.

5. Summary

In this chapter, a detailed description of SM signal generation and detection for VLC systems has been discussed. An overview of different variants of optical SM—OSSK, GSSK, SPPM, S-PAM, and GSM—with their error performance analysis under VLC channel impairments and AWGN has been presented. The analyses of the error performance of these variants have been derived using union bound method and ML criteria. The analytical expressions enabled the theoretical evaluation of the error probability of optical SM technique in a typical MIMO VLC system. Results showed a perfect match between theory and simulations. A summary of the differences between optical SM and other MIMO transmission schemes like SMP, RC, and spatial diversity has also been discussed. The optical SM is thus a technique capable of delivering high data rates in the presence of the limitations of the optical front-end devices. This chapter concludes with the comparison of the spectral and energy efficiency of the variants of optical SM.

Author details

Hammed G. Olanrewaju[1], Funmilayo B. Ogunkoya[2] and Wasiu O. Popoola[1]*

*Address all correspondence to: w.popoola@ed.ac.uk

1 University of Edinburgh, UK

2 Obafemi Awolowo University, Nigeria

References

[1] Mesleh R, Elgala H, Haas H. Optical spatial modulation. Journal of Optical Communications and Networking. 2011;**3**(3):234–244

[2] Mesleh R, Elgala H, Haas H. Spatial modulation. IEEE Transactions on Vehicular Technology. 2008;**57**(4):2228–2241

[3] Ijaz M, Tsonev D, McKendryt JJD, Xiet E, Rajbhandari S, Chun H, et al. Experimental proof-of-concept of optical spatial modulation OFDM using micro LED. In: IEEE Int. Conf. Commun. Work.; IEEE; 2015. pp. 1338–1343

[4] Popoola WO, Poves E, Haas H. Spatial pulse position modulation for optical communications. Journal of Lightwave Technology. 2012;**30**(18):2948–2954

[5] Olanrewaju HG, Thompson J, Popoola WO. Generalized spatial pulse position modulation for optical wireless communications. In: 84th IEEE Vehicular Technology Conference (VTC Fall); IEEE; 2016

[6] Faith T, Di Renzo M, Haas H. On the performance of space shift keying for optical wireless communications. In: IEEE Globecom Workshops; 2010; Miami, FL. IEEE; 2010. pp. 990–994

[7] Fath T, Haas H. Performance comparison of MIMO techniques for optical wireless communications in indoor environments. IEEE Transactions on Communications. 2013;**61**(2):733–742

[8] Popoola WO, Poves E, Haas H. Error performance of generalised space shift keying for indoor visible light communications. IEEE Transactions on Communications. 2013;**61**(5):1968–1976

[9] Fath T, Haas H, Di Renzo M, Mesleh R. Spatial Modulation applied to Optical Wireless Communications in Indoor LOS Environments. In: IEEE Global Telecommunications Conference (GLOBECOM); IEEE; 2011. pp. 1–5

[10] Zhang X, Dimitrov S, Sinanovic S, Haas H. Optimal power allocation in spatial modulation OFDM for visible light communications. In: IEEE 75th Veh. Technol. Conf. (VTC Spring); IEEE; 2012. pp. 1–5

[11] Ghassemlooy Z, Popoola W, Rajbhandari S. Optical Wireless Communications: System and Channel Modelling with MATLAB®. 1st ed. Boca Raton, FL: CRC Press; 2012

[12] Jeganathan J, Ghrayeb A, Szczecinski L, Ceron A. Space shift keying modulation for MIMO channels. IEEE Transactions on Wireless Communications. 2009;**8**(7):3692–3703

[13] Jeganathan J, Ghrayeb A, Szczecinski L. Generalized space shift keying modulation for MIMO channels. In: IEEE 19th Int. Symp. Personal, Indoor and Mobile Radio Commun. (PIMRC); IEEE; 2008. pp. 1–5

[14] Popoola WO, Poves E, Haas H. Demonstration of the merit and limitation of generalised space shift keying for indoor visible light communications. Journal of Lightwave Technology. 2014;**32**(10):1960–1965

[15] Popoola WO, Poves E, Haas H. Generalised space shift keying for visible light communications. In: 8th Int. Symp. Commun. Syst. Networks Digit. Signal Process. (CSNDSP); IEEE; 2012. pp. 1–4

[16] Younis A, Serafimovski N, Mesleh R, Haas H. Generalised spatial modulation. In: 44th Asilomar Conference on Signals, Systems and Computers (ASILOMAR); IEEE; 2010. pp. 1498–1502

[17] Alaka SP, Narasimhan TL, Chockalingam A. Generalized spatial modulation in indoor wireless visible light communication. In: IEEE Global Telecommunications Conference (GLOBECOM); 2015. pp. 1–7

[18] Wang J, Jia S, Song J. Generalised spatial modulation system with multiple active transmit antennas and low complexity detection scheme. IEEE Transactions on Wireless Communications. 2012;**11**(4):1605–1615

[19] Armstrong J. OFDM for optical communications. Journal of Lightwave Technology. 2009;**27**(3):189–204

[20] Barros DJ, Wilson SK, Kahn JM. Comparison of orthogonal frequency-division multiplexing and pulse-amplitude modulation in indoor optical wireless links. IEEE Transactions on Communications. 2012;**60**(1):153–163

[21] Di Renzo M, Haas H, Ghrayeb A, Sugiura S, Hanzo L. Spatial modulation for generalized MIMO: Challenges, opportunities, and implementation. Proceedings of the IEEE. 2014;**102** (1):56–103

[22] Fath T, Haas H. Optical spatial modulation using colour LEDs. In: IEEE International Conference on Communications (ICC); 2013; IEEE; 2013. pp. 3938–3942

[23] Barry JR, Kahn JM, Krause WJ, Lee EA, Messerschmitt DG. Simulation of multipath impulse response for indoor wireless optical channels. IEEE Journal on Selected Areas in Communications. 1993;**11**(3):367–379

[24] Proakis JG, Salehi M. Digital Communications. 5th ed. New York: McGraw-Hill; 2008

Chapter 2

Real-Time Software-Defined Adaptive MIMO Visible Light Communications

Peng Deng

Additional information is available at the end of the chapter

http://dx.doi.org/10.5772/intechopen.68919

Abstract

Visible light communications (VLC) based on light-emitting diodes (LEDs) merges lighting and data communications in applications of Internet-of-Things and 5G networks. However, phosphor-based white LED has a limited linear dynamic range and limited modulation bandwidth. In practical indoor mobile communications, complex channel conditions change dynamically in real-time, and line of sight (LOS) links may be blocked by obstructions. We propose a real-time software-defined adaptive multi-input multi-output (MIMO) VLC system, that both modulation formats (QPSK,16-QAM,64-QAM, 256QAM) and MIMO reconfigurations (Spatial Diversity and Spatial Multiplexing) are dynamically adapted to the changing channel conditions, for enhancing both link reliability and spectral efficiency. Real-time and software defined digital signal processing (DSP) are implemented by Field Programmable Gate Array (FPGA) based Universal Software Radio Peripheral (USRP) devices. We theoretically analysed and experimentally evaluated nonlinear electrical-optical properties and modulation characteristics of white LEDs. We demonstrated a real-time Single-Carrier 256-Quadrature Amplitude Modulation (QAM) 2×2 MIMO VLC, achieving 1.81% averaged error vector magnitude (EVM), 2×10^{-5} bit error rate (BER) after 2 m indoor transmission. As an obstacle moved across LOS links, real-time software-defined adaptive MIMO VLC system enhanced average error-free spectral efficiency of 12 b/s/Hz. This will provide high throughputs for robust links in mobile shadowing environments.

Keywords: visible light communications, light-emitting diodes, multi-input multi-output, spatial multiplexing, spatial diversity, link adaptation, software-defined, channel capacity, nonlinear modulation, equalization, constellation optimization, synchronization

1. Introduction

The world growth in wireless mobile data traffic has led to the development of new technologies for high capacity and energy-efficient wireless communication systems. This fact results in an increasing throughput requirement from the next generation mobile communication networks (5G), which are expected to address several critical challenges, such as broadband capacity, spectral efficiency, power efficiency, quality of services, and mobility coverage. Visible light communication (VLC) system has been accepted as part of the 802.15.7 task group and proposed as a supplement technology in 5G networks standards. VLC based on light-emitting diodes (LEDs) merge lighting and data communications in applications due to their energy efficiency, spectral efficiency, security, and reliability [1]. The white light LEDs have advantages of long lifetime, energy-efficient, and environmental friendliness. VLC using lighting LEDs could provide many advantages, such as no electromagnetic interference (EMI), integration with indoor lighting, available worldwide, and unlicensed bandwidth [2]. The VLC application shown in **Figure 1** is relevant to aerospace, automotive, healthcare industries, 5G networks, wireless personal area networks (WPANs), Internet-of-Things, and ad-hoc networks, as it focuses on the development of radio frequency (RF)-interference-free high-bandwidth communication networks using white light LEDs.

The commercial white light LED utilized for general lighting is mostly based on a blue LED chip covered by a phosphor layer rather than red-green-red (RGB) LEDs, as phosphor white LEDs have much lower complexity and cost. However, phosphor-LEDs have limited modulation bandwidth due to the phosphorescent components. Several modulation bandwidth extension technologies have been investigated, such as optical blue filtering, analogue equalizer, and digital equalization [3]. However, optical filtering of blue light causes large optical intensity loss and decreases visible light transmission range.

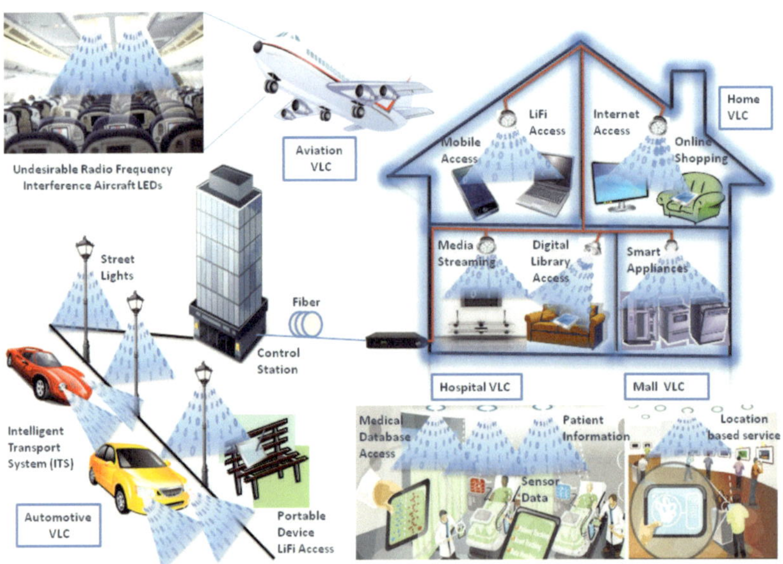

Figure 1. Visible light communication applications.

The capacity of the VLC system under the limited bandwidth can be enhanced by using high spectrally efficient modulation formats, such as orthogonal frequency-division multiplexing (OFDM) [4], Nyquist single-carrier modulation, and carrier-less amplitude and phase (CAP) modulation [5]. Using bit- and power-loading techniques, OFDM can maximize the transmission capacity under the constraint bit error rate (BER). However, high peak-to-average-power ratio (PAPR) of OFDM signal is the crucial challenge for the limited linear dynamic range of white LEDs and nonlinear modulation characteristics in the VLC system [6, 7]. Implementation of real-time CAP systems requires relatively high speed and high-resolution analog to digital converter (ADC), which suffers from high sampling jitter sensitivity. Quadrature amplitude modulation (QAM) architectures have tolerances to nonlinear distortions and sampling jitter. Despite the complexity of adding analog multipliers, QAM systems reduce the sampling rate and resolution requirements of A/Ds for real-time applications.

Moreover, LED-based VLC naturally constitutes a multi-input multi-output (MIMO) system, as large numbers of lighting lamps made up of multiple LEDs are provided by indoor illumination systems [8]. MIMO transmission techniques can be classified by spatial diversity, spatial multiplexing, and spatial modulation. Note that 4×9 spatial multiplex MIMO VLC using 16-QAM OFDM is demonstrated at 250 Mb/s per channel over 1 m range [9]. And 4×4 spatial multiplex MIMO VLC using OOK modulation achieved a data rate of 50 Mb/s over 2 m [10]. Also, 2×2 MIMO VLC using 4-QAM Nyquist single-carrier with frequency domain equalization is demonstrated at 500 Mb/s in 40-cm transmission distance [11]. A wide filed-of-view receiver using fisheye lens is proposed in the spatial-diversity MIMO VLC system [12, 13].

However, most MIMO VLC systems currently manifest the fixed MIMO technology, static channel condition estimation, and offline signal processing. In fact, complex channel conditions are expected to be changed dramatically, and spectral efficiency must be optimized adaptively in practical indoor wireless mobile communications [14, 15]. For example, the MIMO VLC system using fixed spatial multiplexing will suffer link interruption, once any sub-channel condition becomes degraded [16, 17]. Real-time data processing on field programmable gate arrays (FPGAs) based software-defined device can build a fully real-time reconfigurable and an adaptive optical wireless system by upgrading the software [18, 19]. The network architecture should also be designed to support the adaptive MIMO schemes with distributed smart antenna systems [20, 21]. Thus, a real-time adaptive MIMO optical wireless system is necessary to deal with the dynamic complex channel conditions and enhance the robust mobility and link reliability in future wireless mobile networks.

In this chapter, we propose an adaptive MIMO VLC solution that both modulation schemas and MIMO schemas are dynamically adapted to the changing channel conditions for enhancing error performance and spectral efficiency. We experimentally demonstrate a real-time software-defined single-carrier M-QAM MIMO VLC system by using link adaptation of spatial multiplexing and spatial diversity. The VLC transceivers use independent phosphor-white LEDs and PIN photodetectors without blue filtering. We theoretically analyze and experimentally evaluate nonlinear electrical-optical properties and nonlinear modulation characteristics of white LEDs in the VLC system. LED nonlinear modulation characteristics reveal that the optimized LED drive current that corresponds to minimum BER increases with an increase in data rate. We explore FPGA-based universal software radio peripheral devices for real-time digital signal processing. Software-defined implantation of MIMO VLC can enable an adaptive and reconfigurable communication system without hardware changes. We measured the error vector magnitude (EVM), bit error rate (BER), and spectral efficiency performance for single-carried M-QAM MIMO VLC using spatial diversity and spatial multiplexing. Results show that spatial diversity MIMO VLC improves error performance at the cost of spectral efficiency that spatial multiplexing should enhance. With an obstacle moving across line-of-sight (LOS) links, the average error-free spectral efficiency of adaptive 2×2 MIMO VLC achieved 12 b/s/Hz over 2 m indoor dynamic transmission after an obstruction. Real-time software-defined adaptive MIMO VLC will enhance spectral efficiency and link reliability in a mobile environment with shadowing.

2. Nonlinear modulation characteristics of white LED in VLC system

Nonlinear modulation characteristics and limited linear dynamic range of white LEDs are dependent on drive currents, which determine VLC system performance. The static nonlinear LED transfer functions for voltage or optical power to current conversion [22] take into account just clipping effects or continuous-gradient nonlinearity [6]. Moreover, the approximating polynomial model of LED nonlinearity is only valid for small modulation frequencies [6]. Thus, we theoretically analyze a complete nonlinear modulation model based on rate equation [7] and experimentally evaluate the effect of LED drive current on nonlinear modulation characteristics of VLC system [23].

2.1. Nonlinear electrical-optical characteristics of white LED

LEDs are nonlinear devices such as the series resistance and capacitance depend strongly on the bias current. A nonlinear resistor represents the *I-V* characteristic of the LED by the Shockley equation [24]. $i_d = I_0\left[\exp\left(qv_d/nkT\right) - 1\right]$, where k is the Boltzmann's constant, q is the electron charge, T is the junction temperature, n is the diode ideality factor, and v_d is the junction voltage. The factor n is approximately 2 for an InGaN LED. I_0 is the reverse saturation current [24].

However, the LED nonlinear *I-V* characteristic needs to be modified in order to take into account parasitic resistances. A series resistance can be caused by excessive contact resistance or by the resistance of the neutral regions [24]. The equivalent circuit for the transient behavior of LED is shown in **Figure 2**. Considering the nonlinear series resistances R_s and parallel resistance R_p of LED, the *I-V* characteristic of a forward-biased p-n junction diode ($v_d = v\text{-}IR_s$, $v_d \gg (kT/q)$ is given by $I - (v - IR_s)/R_p = I_0 \exp\left[q(v - IR_s)/(nkT)\right]$. For devices with a high parallel resistance ($Rp \to \infty$), the diode *I-V* characteristic can be written as

$$I = I_0 \exp\left[e(v - IR_s)/(nkT)\right] \tag{1}$$

http://dx.doi.org/10.5772/intechopen.68919

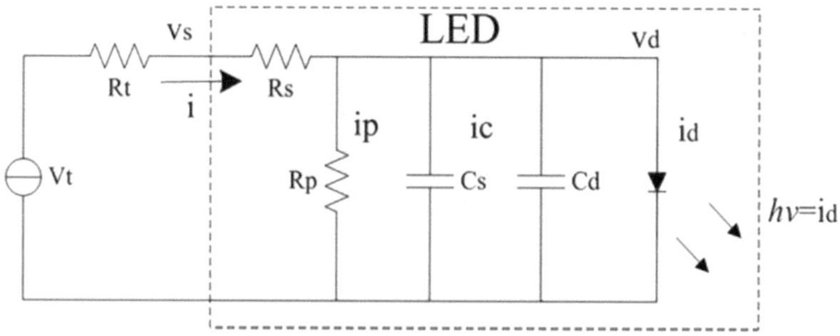

Figure 2. The equivalent circuit for the transient behavior of LED.

For $Rp \rightarrow \infty$ and $Rs \rightarrow 0$, this equation reduces to the Shockley equation.

We measured electrical-optical properties of LED chips. **Figure 3(a)** shows experimental and theoretical relationship between drive current I against forward voltage v for a single LED. The measured LED bias current I against the junction voltage v_d characteristic well matches the theoretical nonlinear I-v_d model. **Figure 3(b)** shows relationship between optical power and forward voltage for a single LED. LED optical power get cut off below the threshold of 10 mA @ 2.7 V, increased within a linear dynamic range of 600 mV around bias point of 450 mA @ 3.3 V, finally saturated and decreased beyond 1100 mA @ 3.7 V. This is because nonlinear thermal effects appear to reduce the internal quantum efficiency and electrical to optical conversion efficiency. Thus, the LED drive current and linear dynamical range should be optimally designed to avoid nonlinear distortions in output optical power and degradations in VLC performance [7].

Figure 4 shows LED bias current and output optical power against forward voltage for two parallel-connected and series-connected LED chips. It can be seen that the nonlinear region of parallel-connected LED chips fluctuates dramatically above 3000 mA @ 3.5 V due to the larger forward current thermal effect, whereas the linear modulation dynamic range of the series-connected LED chips increases to 1200 mV around the optimal DC-bias point 370 mA @ 6.4V due to the improved impedance. Thus, series-connected LED chips can enhance the linear dynamic range.

We evaluated VLC transceiver system response at the frequency of 10 MHz along the distance of 1 m. Using the same transmitter signal amplitude of 100 mV, we measured the receive signal amplitude for different LED DC-bias power levels. **Figure 5** shows the effect of LED forward voltage and current on the AC signal system response for two and three series-connected LED chips. It is shown that the received signal and system response fluctuate to decline gradually with the increase of LED DC-bias current and voltage, because high DC power will lead to a nonlinear thermal effect that saturated output optical power and degraded AC signal amplitude. As compared, a larger number of series-connected LED chips can mitigate the nonlinear fluctuation of system response.

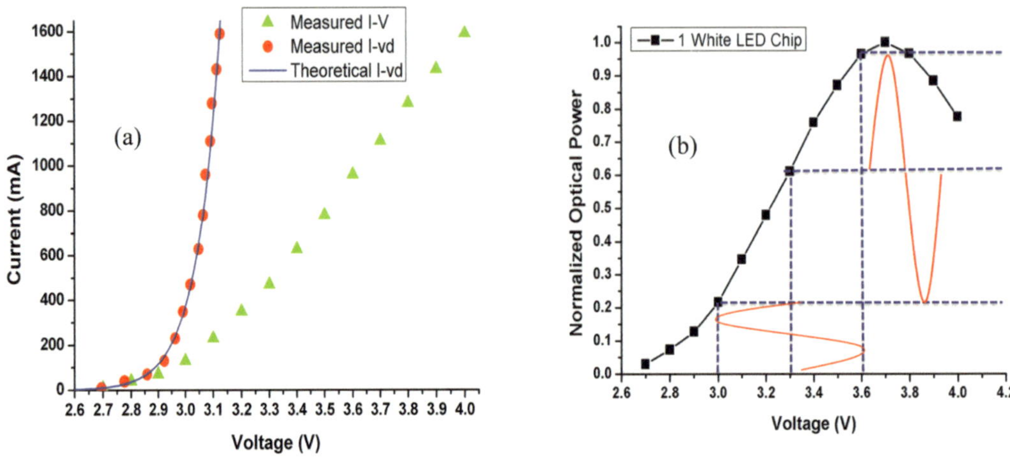

Figure 3. Electrical to optical properties of white LED. (a) DC-bias current against forward voltage and (b) normalized optical power against forward voltage for single LED.

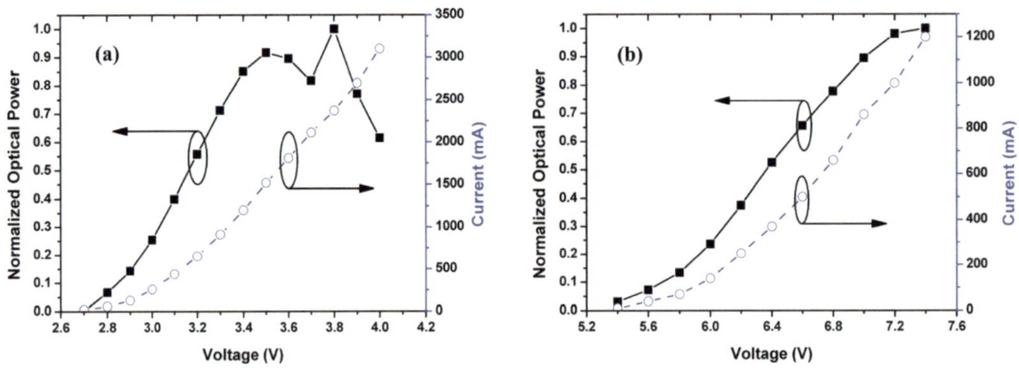

Figure 4. DC-bias current and normalized optical power against forward voltage for (a) two parallel connected LEDs and (b) two series-connected LEDs.

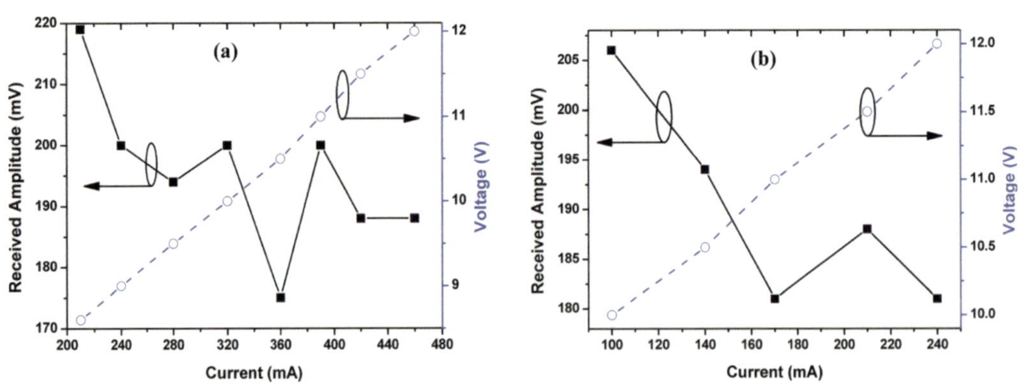

Figure 5. DC-bias current and voltage effect on VLC system response for (a) two series-connected LED chips and (b) three series-connected LED chips. Transmitter signal amplitude is 100 mV.

2.2. Nonlinear modulation bandwidth of white LED

2.2.1. LED modulation bandwidth due to carrier lifetime

The differential carrier lifetimes in InGaN-based LEDs can be described by the simple ABC model. This rate equation model considers that the current through the device is made up of three contributions: a nonradiative current I_A due to Shockley-Read-Hall (SRH) recombination at defect sites, a current I_B due to radiative recombination of electrons and holes, and an Auger current I_C due to cubic nonradiative recombination [25]. The total current is thus

$$I = I_A + I_B + I_C = qsd(\mathrm{AN} + \mathrm{BN}^2 + \mathrm{CN}^3) \tag{2}$$

where I is the operation current, q is the elementary charge, s and d are the device area and total thickness of the quantum wells (QWs), N is the carrier density, A, B, and C are the coefficients of Shockley-Read-Hall (SRH), radiative recombination and Auger nonradiative recombination.

Under small-signal modulation conditions, the luminescence decay is mono-exponential with the differential time constant [26]. The differential lifetime is then given by the derivative of the recombination rate with respect to carrier density $1/\tau_s = A + \mathrm{BN} + \mathrm{CN}^2$. In the case that linear and quadratic terms dominate the recombination rate, the simplified expression between carrier lifetime and bias current can be derived by $1/\tau_s^2 = A^2 + 4BI/qV$. Thus, the LED electrical-optical 3-dB bandwidth [23] related to the differential lifetime can be obtained

$$f_s = \frac{1}{2\pi\tau_s} = \frac{1}{2\pi}\sqrt{A^2 + \frac{4B}{qV}I} \tag{3}$$

2.2.2. LED modulation bandwidth due to space-charge capacitance

The common feature of LED is that the large junction over the entire wafer presents a space-charge capacitance of the junction in parallel with the diffusion capacitance of carrier lifetime. **Figure 2** shows the equivalent circuit for the transient behavior of LED. In addition to the diffusion capacitance C_d of carrier lifetime, there is the space-charge capacitance Cs of the depletion layer. The space-charge capacitance [27] is given by $C_s = C_0/(1 - v_d/\phi)^m$, where $C_0 = sc_0$, c_0 is the zero-bias capacitance per unit area, $\varnothing$ is the barrier voltage, and the exponent m is 1/2. The LED resistance R_s in the equivalent circuit accounts for the series resistance of the bulk material and the contact resistances. Thus, the time constant due to the space-charge capacitance [23] is given by $\tau_c = C_sR_s = R_sC_0/(1 - v_d/\phi)^m$. The electrical-optical 3-dB bandwidth related to space-charge capacitance can be obtained

$$f_c = \frac{1}{2\pi\tau_c} = \frac{(1 - v_d/\phi)^m}{2\pi R_s C_0} \tag{4}$$

2.2.3. Nonlinear modulation bandwidth of white LED

Considering the nonlinear electrical-optical characteristics of LED, substituting Eq. (1) into Eq. (4), the total LED rise time due to carrier spontaneous lifetime and space-charge capacitance [23] can be defined by

$$\tau_{\mathrm{LED}}(I) = \tau_s + \tau_c = \frac{1}{\sqrt{A^2 + 4BI/qV}} + \frac{R_s C_0}{\sqrt{1 - \ln(I/I_0)nkT/q\phi}} \tag{5}$$

The total equivalent electrical-optical 3-dB bandwidth due to the differential lifetime and space-charge capacitance can be expressed by the LED bias current

$$f_{\mathrm{LED}}(I) = \frac{1}{2\pi(\tau_s + \tau_c)} = \frac{1}{2\pi} \frac{\sqrt{(A^2 + 4BI/qV)[1 - \ln(I/I_0)nkT/q\phi]}}{\sqrt{1 - \ln(I/I_0)nkT/q\phi} + R_s C_0 \sqrt{A^2 + 4BI/qV}} \tag{6}$$

We measured the optical spectrum of different phosphor white LED chips (Osram Ostar) as shown in **Figure 6**. It can be seen that a commercial warm white LED (CWLED) contains only 10% blue component of the overall emitted power, while the emitted ultra-white LED light (UWLED) consists of 41% blue component and less slow yellow component from the phosphor. As the modulation bandwidth of blue light is higher than that of yellow component, the ultra-white LED (UWLED) can be used to improve the modulation bandwidth efficiency [7].

The effect of LED DC-bias current on electrical-optical-electrical (EOE) channel frequency response of the VLC transceiver is shown in **Figure 7**. The magnitude of the channel frequency response is measured by a network analyzer. It can be seen that the magnitude of frequency response reduced when bias current is less than 100 mA due to clipping of the lower peaks at the turn on point. For small frequency signals less than 10 MHz, VLC system frequency response

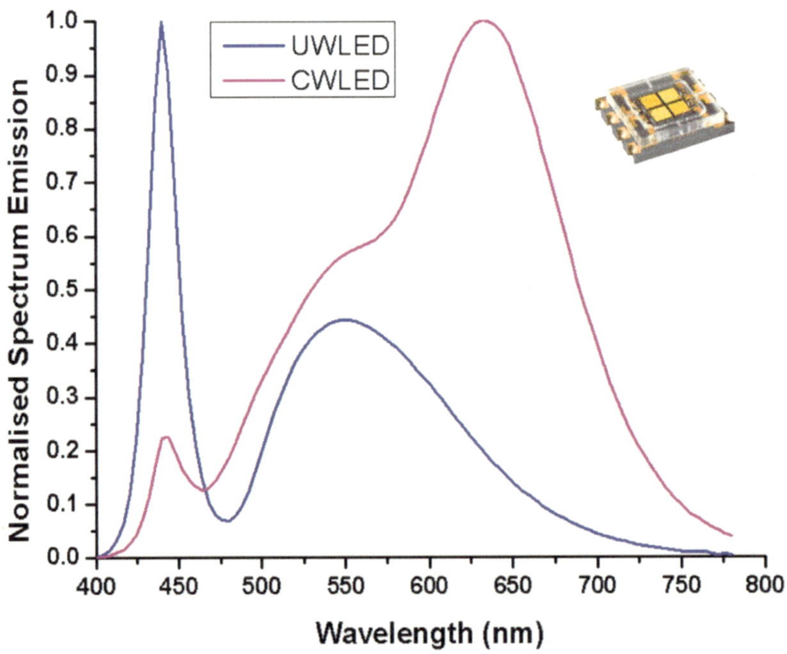

Figure 6. The measured optical spectrum of different white LED chips, inset: Osram white LED.

magnitudes reduce as LED bias currents increase, due to nonlinear intensity saturation at large currents. However, for high frequency signals in **Figure 7(b)**, the system response improves with the increasing DC bias current due to higher optical power detected at the receiver [7].

The comparison of experimental and theoretical relationship between 3-dB bandwidth of the VLC system and LED bias current is shown in **Figure 8**. It is found that the 3-dB bandwidth of the VLC system increases with LED bias current, because the bandwidth due to carrier lifetime f_S in Eq. (3) increases significantly with carrier density and driving current. Furthermore, the LED bandwidth saturates and approaches to 7 MHz when DC bias current is larger than 250 mA, because the bandwidth due to space-charge capacitance f_c in Eq. (4) reduces apparently with bias current. It is shown that results of experimental measurement of LED bias current-related equivalent bandwidth well match the theoretical results in Eq. (6). Since the receiver itself has a relatively large bandwidth (photodetector 150 MHz, amplifier 1 GHz), the system bandwidth (7 MHz) is limited by the LED module.

As the blue filtering introduces high optical power attenuation, we extend LED modulation bandwidth without blue filtering. The magnitude of the channel frequency response was measured by a network analyzer. A first-order equalizer that consists of a capacitor in parallel with a resistor ($R = 1$ kΩ, $C = 30$ pF) is developed to enhance the system bandwidth. **Figure 9** presents the frequency response of the equalizer and the corresponding equalized VLC transceiver system. It can be seen that the optimized analogue equalizer can extend the bandwidth of the VLC transceiver from 7 to 40 MHz.

2.3. Nonlinear modulation effects of white LED on VLC performance

Figure 10 presents a diagram of experimental measurement setup for the VLC system. The analogue transmitter front-end superimposed the amplified output signal onto LED bias current using a bias-tee. The phosphorescent white LED (Osram OSTAR) with four chips provided 520-lm luminous flux with a 120° opening angle at 700 mA current. An aspheric lens collimated LED light propagation along line-of-sight paths. A high-speed PIN photodetector captured LED light by an aspheric convex lens. A low noise transimpedance amplifier (TIA)

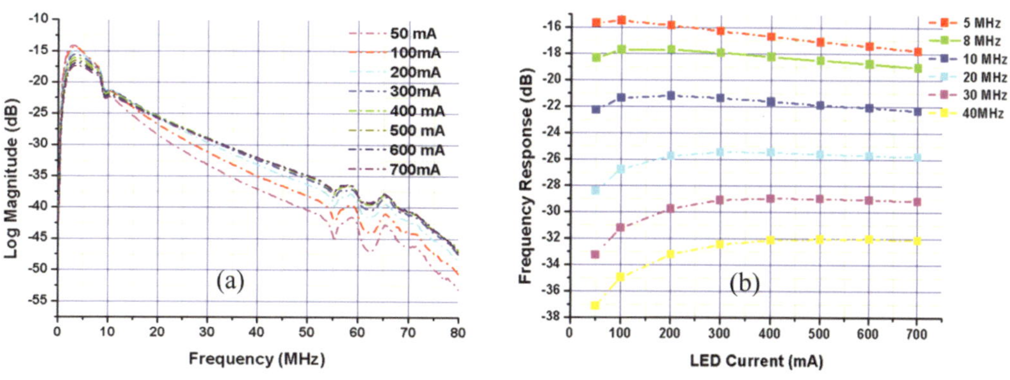

Figure 7. (a) The measured EOE channel frequency response of VLC transceiver for different LED DC bias currents and (b) frequency response versus LED DC bias current for different frequency.

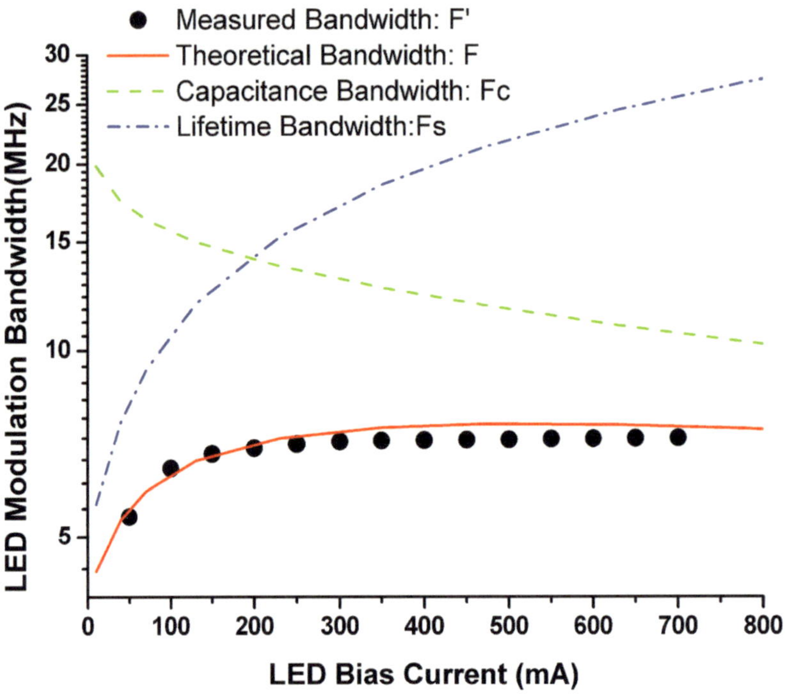

Figure 8. Relationship between 3-dB bandwidth and average LED current.

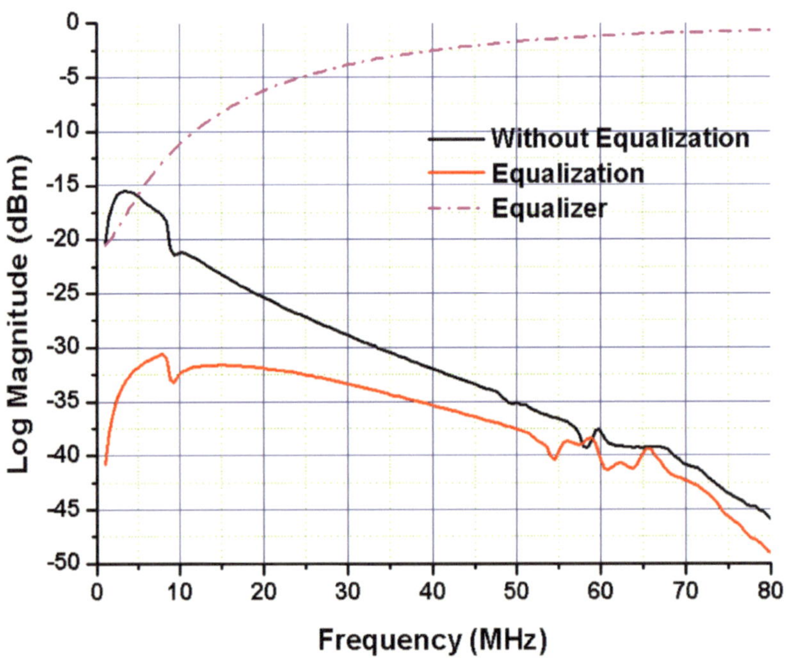

Figure 9. Electrical-optical-electrical system frequency response of VLC transceiver without equalization, VLC transceiver with equalization and the analogue equalizer.

amplified the photocurrent signal, which bandwidth was extended by a first-order analogue postequalizer [16]. The received electrical signal was amplified and then connected to a digital oscilloscope (HP 4396B) and a bit error rate tester (BERT).

A pseudorandom binary sequence (PRBS)-10 OOK-NRZ data stream with a peak-to-peak voltage swing of 2.5 V is used to modulate a single LED light. **Figure 11** shows the relationship between the measured bit-error-rate (BER) performance and the date rate of the equalized VLC system for different DC bias currents. It can be seen that the maximum date rate at DC bias current of 510 mA can achieve 60 Mbps for a BER threshold of 10^{-3}. In contrast, BER performance degraded obviously for both a low DC bias current of 200 mA and a high DC bias current of 680 mA.

The effect of LED DC bias current on VLC BER performance for different data rates is illustrated in **Figure 12**. For a BER threshold of 10^{-3}, the maximum date rate of 60 Mbps can be achieved at 510 mA bias current. The smaller bias currents less than 100 mA increased the BER at low data rates (20 and 32 Mbps). The DC bias current that corresponds to minimal BER value is used to identify the optimum bias point, where the LED is operating in a linear region with less probability of clipping the signal peaks. The high BER observed at the large bias current more than 200 mA is mainly due to the nonlinear thermal effect and the clipping of the upper peak at the maximum output optical power [7]. For high data rates (40 and 60 Mbps), BER decreases to the minimum value with an increase in LED bias current. This is because relatively high levels of optical power are required to overcome the received signal-to-noise penalty at a high data rate. Therefore, the optimized DC-bias current of an LED driver that corresponds to minimum BER increases gradually with an increase in the data rate. The modulation bandwidth can be enhanced by reducing parasitic resistances and capacitances of LED, or carrier sweep-out of the active region. This can help in controlling the dimmable illumination brightness of white LEDs to match the modulation speed and improve system performance.

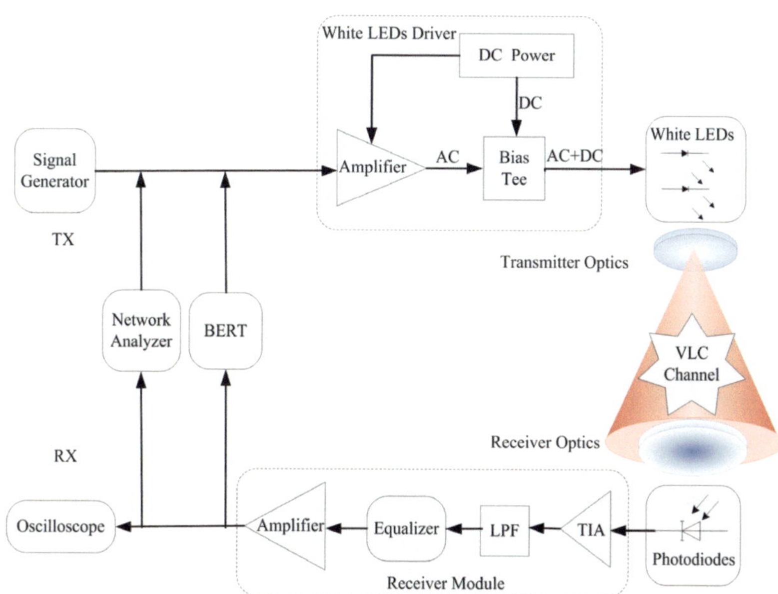

Figure 10. Experimental setup of visible light communication system using white LEDs.

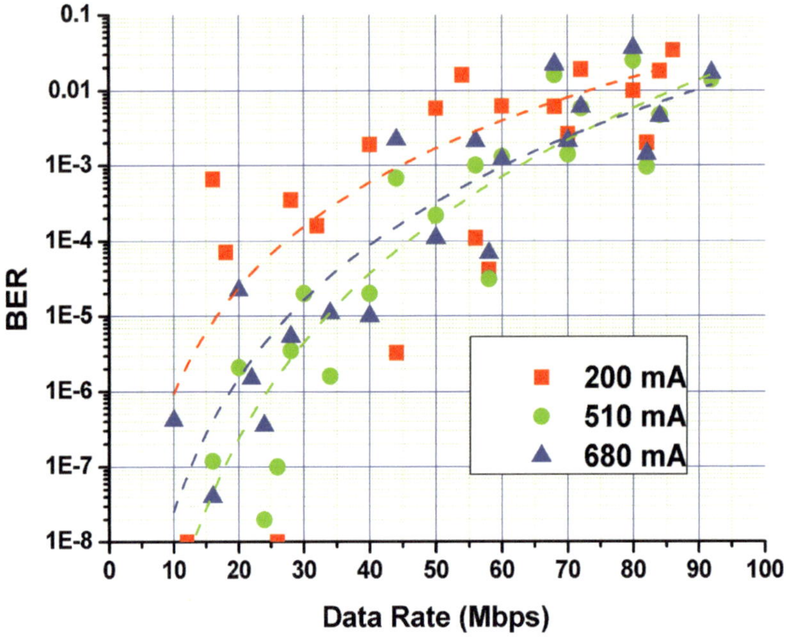

Figure 11. Relationship between VLC bit error rate and date rate for different LED DC-bias currents.

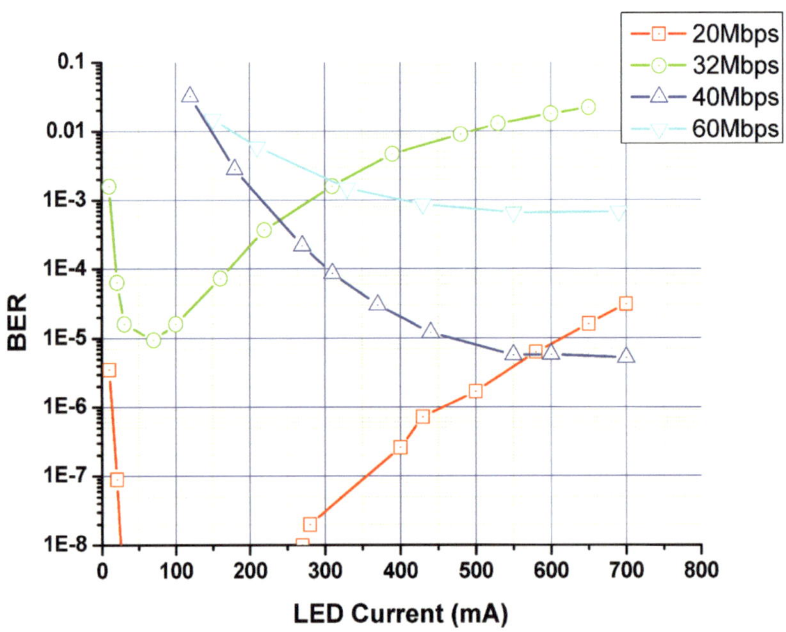

Figure 12. Relationship between VLC bit error rate and LED current for different date rate.

3. Real-time software-defined adaptive MIMO VLC system

3.1. System model of adaptive MIMO VLC

A MIMO visible light communication system consists of M transmitting antennas (LEDs) and N receiving antennas (photodetectors). The input data X are mapped by QAM and divided into M parallel symbol streams x_j ($j = 1,\ldots, M$) to independently modulate a light source. The transmitted signal after up-conversion takes the form [16]:

$$x(t) = I(t)\cos(2\pi f_c t) - Q(t)\sin(2\pi f_c t) \tag{7}$$

where f_c is the carrier frequency, $I(t)$ and $Q(t)$ are the modulating signals of in phase and quadrature components. The received signal is a linear combination of all x_j, and the MIMO channel model can be represented as [16]

$$\mathbf{y} = \mathbf{H}\mathbf{x} + \mathbf{n}, \tag{8}$$

where $\mathbf{n}$ is the sum of thermal noise and ambient shot light noise with zero mean and a variance σ^2, and the noise power spectral density is N_0. It is assumed that N_t data symbols x_1, x_2, …, x_{Nt} are chosen randomly, equally likely and independently to form an input data vector $\mathbf{x} = [x_1, x_2, \ldots, x_{N_t}]^T \in \mathbf{A}^M$ where $\mathbf{A}$ is a given modulation constellation. Each Tx transmits independent symbols of the same power: $<\mathbf{x}\mathbf{x}^*> = P_t/N_t.\mathbf{I}$, where P_t is the total Tx power, $\mathbf{I}$ is the identity matrix, <> and * denote the expectation and Hermitian conjugation, respectively. The received signal vector is denoted by $\mathbf{y} = [y_1, y_2, \ldots, y_{N_r}]^T$. Notation $\mathbf{H}$ denotes an $N_t \times N_r$ channel matrix, in which h_{nm} is the real nonnegative coefficient between the nth receiver and the mth transmitter.

Channel gain of an indoor VLC LOS link is determined by [28]

$$h_{nm} = \begin{cases} \dfrac{(k+1)A}{2\pi d_{nm}^2}\cos^k(\phi)\cos(\psi), 0 \le \psi \le \Psi_{\frac{1}{2}} \\ 0, \qquad\qquad\qquad\qquad\qquad \psi > \Psi_{\frac{1}{2}} \end{cases} \tag{9}$$

with ψ is the angle of incidence with respect to the nth receiver axis, and φ is the angle of emergence with respect to the mth transmitter axis. d_{nm} is the distance between mth transmitter and nth receiver. Ψ is the field-of-view (FOV) of the receiver, A is the detector area, and $k = -\ln 2/\ln(\cos(\varphi_{1/2}))$, in which $\varphi_{1/2}$ is the half power angle of the transmitter.

The MIMO system has first to estimate channel estimation coefficients between Tx and Rx, to de-multiplex signals and retrieve transmitted data. We periodically inserted training sequences of binary phase shift keying signals in front of the data streams to obtain the channel estimation matrix. Each row of the two-dimensional (2D) array contains two training sequences, one from each transmitter [16].

$$\mathbf{t} = \begin{pmatrix} ts_1 & 0 \\ 0 & ts_2 \end{pmatrix} \tag{10}$$

Thus, we can determine the four channel estimates by performing channel estimation on each training sequence in each row. An **H** matrix detailing channel estimation is logged on receiving all pilot signals $\mathbf{y_t}$.

$$\mathbf{H} = \frac{\mathbf{y_t}}{\mathbf{t}} = \begin{pmatrix} y_{11}/ts_1 & y_{12}/ts_2 \\ y_{21}/ts_1 & y_{22}/ts_2 \end{pmatrix} \tag{11}$$

Zero-forcing (ZF) is the simplest method to estimate the transmitted data, that invert **H** and multiply it with the received vector [29]: $\mathbf{W}^*\mathbf{y} = \mathbf{x}_{\text{est}} + \mathbf{n}$, where W is the beam former $\mathbf{H}^{-1}$. However, if **H** is rank deficient, matrix inversion cannot be performed. The noise vector **n** will lead to noise amplification for low values of **H**. Thus, the pseudo inverse of **H** can be used by [16]:

$$\mathbf{H}^{\dagger} = (\mathbf{H}^*\mathbf{H})^{-1}\mathbf{H}^*, \tag{12}$$

where $\mathbf{H}^*$ is the conjugated transpose of **H**. Using the pseudo inverse zero-forcing, the de-multiplexing and postequalization can be simultaneously realized for an estimate of **x**.

The adaptive MIMO is mainly based on the approximated expression for the average bit error rate (BER) of M-ary QAM symbols and channel capacity of the MIMO system. An approximate expression of average bit error probability of M-ary square QAM can be obtained by

$$P_b \cong \frac{2\left(\sqrt{M}-1\right)}{\sqrt{M}\log_2^M} erfc\left(\sqrt{\frac{3}{2(M-1)}\gamma}\right) \tag{13}$$

where γ is the received signal to noise ratio (SNR) per symbol, and $erfc(x)$ is the complementary error function. The analytical error-probability bound for an AWGN SISO channel with M-QAM modulation is given by [30]

$$P_b \leq \frac{1}{5}\exp\left(-1.5\frac{\gamma}{M-1}\right) \tag{14}$$

Assuming that signals from individual LEDs are independently equipowered, and a channel matrix is perfectly known to the receiver [29], spectral efficiency of MIMO channels is obtained by

$$\begin{aligned} S &= \log_2\left[\det\left(\mathbf{I}_N + \frac{P_t}{N_t\sigma_0^2}\mathbf{H}\mathbf{H}^*\right)\right] \\ &= \sum_{i=1}^{N_m}\log_2\left(1 + \frac{P_t}{N_t\sigma_0^2}\lambda_i^2\right) \\ &\leq \log_2\left(1 + \frac{P_t}{N_t\sigma_0^2}\|\mathbf{H}\|^2\right) \end{aligned} \tag{15}$$

where $N_m = \min(N_t, N_r)$, $\boldsymbol{\lambda} \triangleq \left[\lambda_1^2, \ldots, \lambda_{N_m}^2\right]^T$ is eigenvalue vector of $\mathbf{H}\mathbf{H}^*$, $\|\|$ denotes the Frobenius norm, $\|\mathbf{H}\|^2 = \sum_{i=1}^{N}\sum_{j=1}^{M}|h_{i,j}|^2$.

For a given BER_{tgt}, the maximum spectral efficiency can be obtained by

$$s = \log_2(1 + K\gamma) \tag{16}$$

where $K = \frac{-3}{2\ln(5BER_{tgt})}$.

The adaptive MIMO system can be defined by the constrained optimization problem.

$$\max_{\text{MIMO}, s_i, P_i} S = E_{\boldsymbol{\lambda}}\left[\sum_{i=1}^{N_m} s_i(\boldsymbol{\lambda})\right] \tag{17}$$

subject to

$$E_{\boldsymbol{\lambda}}\left[\sum_{i=1}^{N_m} P_i(\boldsymbol{\lambda})\right] \le P_t \tag{18}$$

$$\text{BER}_\text{i}(\boldsymbol{\lambda}) \le \text{BER}_{tgt} \tag{19}$$

$$M_i(\boldsymbol{\lambda}) \in \{M_0, M_1, \ldots, M_n\} \tag{20}$$

$$\text{MIMO}(\boldsymbol{\lambda}) \in \{SM, SD\} \tag{21}$$

$$P_i(\boldsymbol{\lambda}) \ge 0, s_i(\boldsymbol{\lambda}) \ge 0, i = 1, \ldots, N_m \tag{22}$$

Eq. (17) expresses the optimization strategy that maximizing the spectral efficiency by adapting the MIMO scheme, transmit rate s_i and power P_i in each subchannel. Eq. (18) shows that the average total transmit power cannot exceed P_t, and Eq. (19) requires that at any instant the BER in any subchannel must remain below a predetermined target level BER_{tgt}. We consider more practical discrete rate systems in Eq. (20), where we restrict the available constellation size M_i to a finite set of integers instead of any nonnegative real values. In particular, we use only square QAMs as the available component modulation schemes; i.e., $M_i = 4^i, i = 0, 1, \ldots, N$. Based on the conditions of MIMO channel estimations, we can adapt the MIMO schemes in Eq. (21) such as spatial multiplexing (SM) and spatial diversity (SD) to maximize the achievable spectral efficiency. Eq. (22) is the nonnegative constraint on power and rate.The expectation over $\boldsymbol{\lambda}$ in Eq. (17) implies that we assume the random process $\boldsymbol{\lambda}$ is ergodic. These enable the MIMO adaptation in the time domain and spatial domain.

3.2. Experimental setup of adaptive MIMO VLC system

Figure 13 presents a block diagram and experimental setup of this 2 × 2 M-QAM adaptive MIMO VLC system. The random binary data is generated in Labview and would be first split into two parallel streams, one for each transmitter (TX) channel. We develop adaptive MIMO modes control modules, which can adjust the optimal modulation formats and MIMO schemes to maximize the spectral efficiency and error performance according to the real-time MIMO channel conditions. We adaptively choose one of MIMO technologies: spatial diversity (SD) and spatial multiplexing (SM). SD MIMO transmits the same data stream in each Tx to improve antenna array gain, while SM MIMO sends different parallel streams over each LED to enhance spectral efficiency gain. Bit stream is mapped into *M*-ary QAM in each channel, and four types of modulation formats are provided: 4-QAM, 16-QAM, 64-QAM, and 256-QAM. Thus, there are eight adaptive MIMO modes available for link adaptation (SM-4, SM-16, SM-64, SM-256, SD-4, SD-16, SD-64, and SD-256). Initially, we choose the default mode of SM-64 (spatial multiplexing 64-QAM). Then channel-training sequences are inserted into the symbol streams. After adding cyclic prefix (CP) and upsampling, pulse shaping by a raised-cosine filter is employed.

Data streams are transmitted from the host computer to USRP X310 using 10G SFP+ Ethernet interfaces. The complex QAM streams are upconverted after digital to analog conversion (DAC), and generate the real analogue RF signals. The analogue transmitter front end superimposed the amplified output signal onto LED bias current using a bias tee. The phosphorescent white LED (Osram OSTAR) with four chips provided 520-lm luminous flux with a 120° opening angle at 700 mA current. An aspheric lens collimated LED light propagation along line-of-sight direction. A high-speed PIN photodetector captured LED light by an aspheric convex lens. A low noise transimpedance amplifier (TIA) amplified the photocurrent signal, which bandwidth was extended by a first-order analogue postequalizer.

Subsequently, we routed the received signals from each receiver to an RX channel of USRP X310. USRP X310 is a scalable software-defined radio (SDR) platform for designing and deploying the next generation wireless communication system. The hardware architecture integrates two daughterboard slots covering DC-6 GHz with up to 120 MS/s of baseband bandwidth, 200 MS/s ADC, and 800 MS/s DAC, and dual high-speed SFP(+) interface ports for 1/10 Gigabit Ethernet. We set 10 MHz baseband bandwidth due to the limited sampling

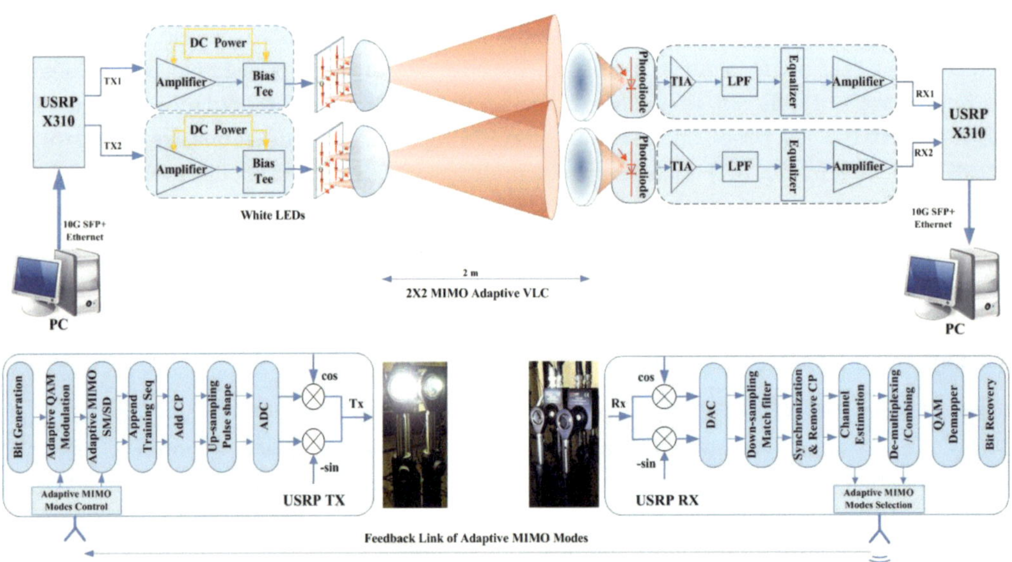

Figure 13. Experiment setup for adaptive MIMO VLC.

rate of daughterboard and analog to digital conversion (ADC). A large user-programmable Kintex-7 FPGA in a convenient desktop provides best-in-class real-time hardware performance [16].

Training sequences are acquired for synchronization and channel estimation, after downconverting to baseband and down sampling by a matched filter. With removing CP, the received data streams are processed in an MIMO de-multiplexer. The final streams are then passed through a QAM demodulator to recover original binary streams. Based on the channel estimation conditions, the adaptive MIMO mode selection module will calculate and select the corresponding optimum modes including modulation formats and MIMO schemes. We then update the three-bit binary code for adaptive MIMO modes and feedback the adaptive mode code to the transmitter by using RF uplink. In Labview, we measured and analyzed the constellation diagram, EVM, BER, and spectral efficiency performance for adaptive *M*-QAM MIMO VLC system.

4. Results and discussion

In order to achieve adaptive MIMO visible light communication for indoor lighting environment, we first evaluate efficient estimation approaches to acquire the real-time channel conditions. Then we measured and compared the error performance and spectral efficiency of M-QAM MIMO VLC systems using spatial diversity and spatial multiplexing, respectively. Based on the adaptive MIMO mode selection criteria in terms of channel estimation, we demonstrated the adaptive real-time software-defined MIMO VLC along with different distance.

4.1. Real-time single-carrier QAM MIMO visible light communication

Figure 14(a) shows the captured waveforms and resulting IQ symbols of RX1 and RX2 for 64-QAM spatial diversity MIMO VLC. From this figure, we can find the crosstalk and phase offset between these two RXs, which require MIMO processing in the DSP. The constellation plots show the signals received by the two Rx NI USRP transceivers.

We recover 2D array of symbols by postequalization with channel estimations and convert them into the proper 1D array of symbols by performing max ratio combining (MRC) as shown in **Figure 6(b)**.

Figure 14(c) and **(d)** shows the recovered signal constellation diagram after channel equalization. **Figure 14(e)** shows the final MIMO combined signal constellation plot after performing max ratio combing (MRC). The measured root mean square (RMS) EVM and BER are 1.73% and zero, respectively. This reconstructed constellation shows that MIMO system works properly.

We then demonstrate spatial multiplexing MIMO VLC by transmitting different data streams. **Figure 15(a)** shows four received I and Q waveforms from each of the two Rx antennas after downsampling and postequalization, which are combined together by MIMO de-multiplexing to recover the MIMO RX symbols IQ graph in **Figure 15(b)**. From **Figure 14(c)–(e)**, we can find that all the 64-QAM IQ symbol waveforms have been successfully reconstructed in the spatial multiplexing MIMO VLC system. Furthermore, we measured high-order modulation performance for 256-QAM spatial multiplexing MIMO VLC, which constellation diagrams are depicted in **Figure 16**. The results show clear constellations and error free transmission in 256-QAM spatial multiplex 2 × 2 MIMO VLC with EVM 1.81%, BER 2×10^{-5} and data rate 160 Mbps along 2 m.

Figure 17 shows comparison of BER versus signal power of a 64-QAM 2 × 2 MIMO VLC transmission. Compared with spatial multiplexing, the spatial diversity MIMO VLC system can enhance about 3-dB array gain in BER performance, as the array gain in spatial diversity MIMO VLC improves the received signal strength.

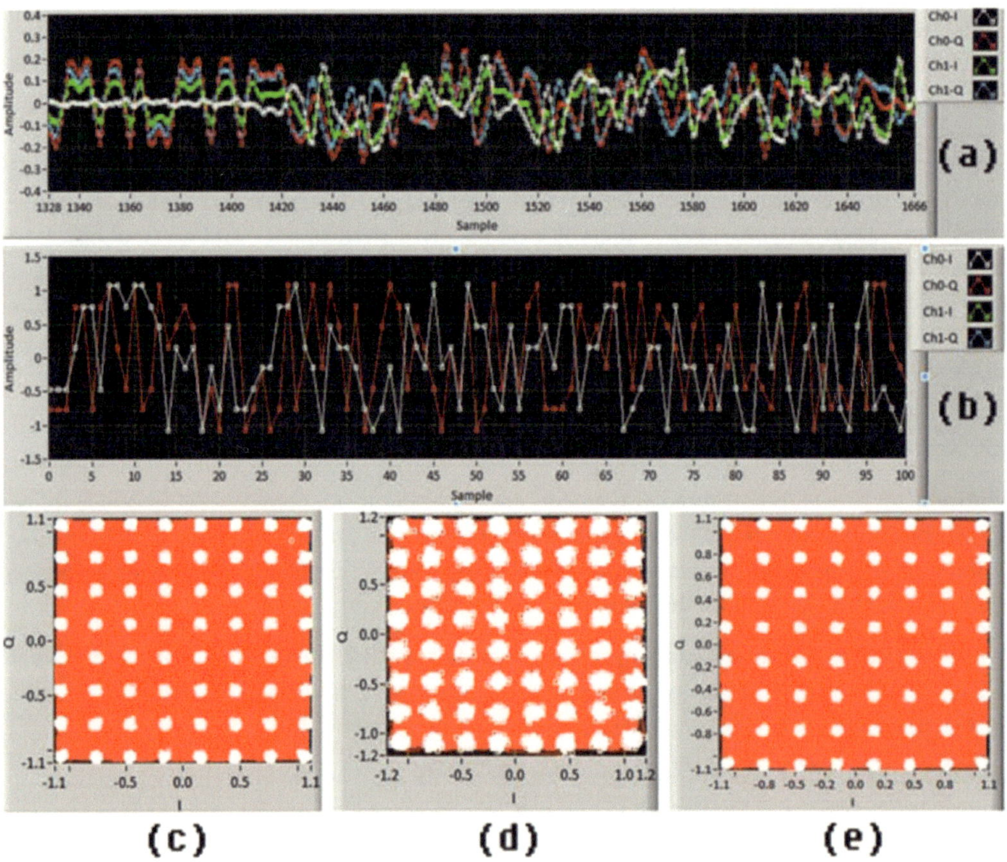

Figure 14. Spatial diversity 64-QAM 2 × 2 MIMO VLC. (a) Rx waveform IQ graph, (b) RX MIMO MRC combined symbol IQ graph, (c) RX CH0 constellation diagram, (d) RX CH1 constellation diagram, and (e) MIMO combined signal after max ratio combing diagram.

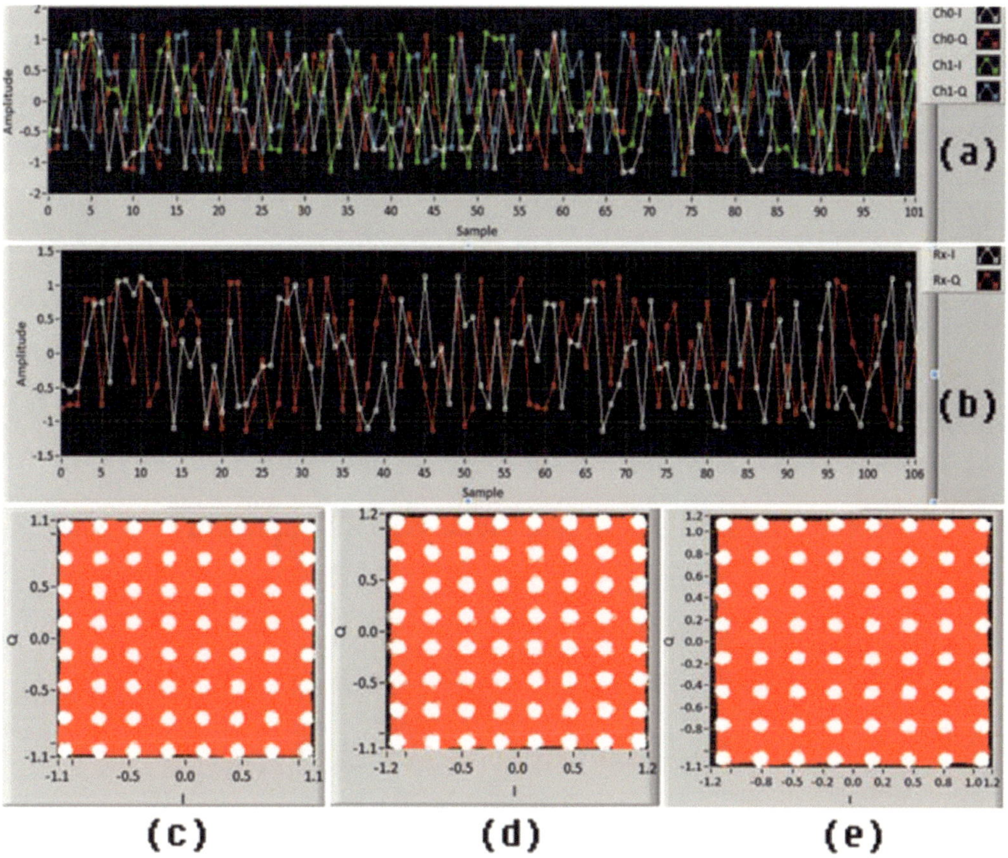

Figure 15. Spatial multiplexing 64-QAM 2 × 2 MIMO VLC. (a) Rx waveform IQ graph, (b) RX MIMO IQ graph, (c) RX CH0 constellation diagram, (d) RX CH1 constellation diagram, and (e) Rx MIMO spatial multiplexing diagram, EVM = 1.89%, BER = 0.

4.2. Software-defined MIMO VLC using link adaptation of spatial diversity and spatial multiplexing

Figure 18 shows the measured EVM and channel response estimation against signal power for 2 × 2 MIMO 4-QAM VLC using spatial multiplexing and spatial diversity. For large symbol streams T, such that $T \gg N$, where N is the number of unique modulation symbols, channel SNR can be obtained from EVM as SNR ≈ $1/\text{EVM}^2$. As shown in **Figure 10**, EVM of SM and SD MIMO VLC decreases correspondingly with an increase of signal power. As EVM approaches to small value of 2%, both EVMs become saturated due to the limited symbols streams. It is clear that EVM of spatial diversity system has a power gain of about 3 dB compared to the spatial multiplexing due to the antenna array gain. Channel SNR estimation from the measured EVM is dependent on the symbol number, constellation size, and MIMO processing. On the other hand, we obtain channel SNR directly from a channel response estimation matrix,

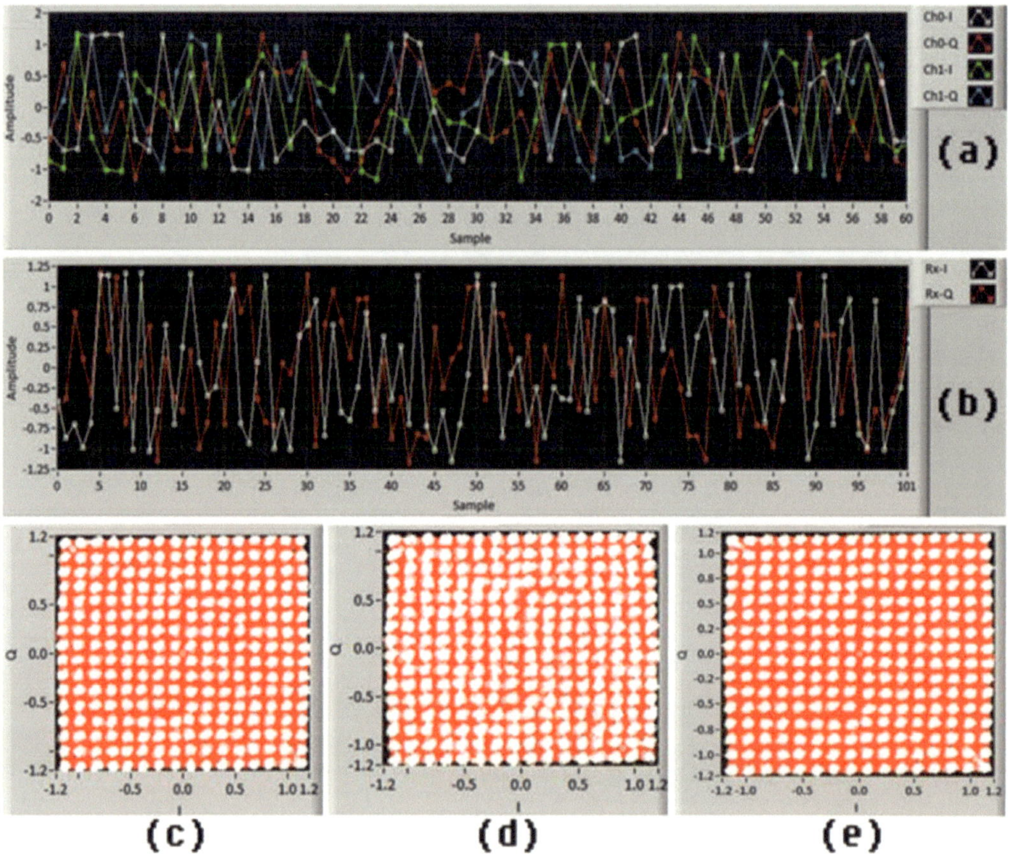

Figure 16. Spatial multiplexing 256-QAM 2 × 2 MIMO VLC. (a) Rx waveform IQ graph, (b) RX MIMO IQ graph, (c) RX CH0 constellation diagram, (d) RX CH1 constellation diagram, and (e) Rx MIMO spatial multiplexing diagram, EVM = 1.81%, BER = 2E-5.

where λ_i is eigenvalue of channel estimation matrix H. As **Figure 10** shows, each subchannel eigenvalue of channel estimation has a linear relationship with signal power. Furthermore, original channel estimation matrix is similar for both spatial diversity and spatial multiplexing MIMO. Thus, we can acquire real-time subchannel SNRs from eigenvalues of channel estimation matrix regardless of MIMO schemes and modulation formats.

BER performance of 4-16-64-256 QAM 2 × 2 MIMO VLC using spatial multiplexing and spatial diversity is shown in **Figure 19**, where dots are experimental measurements and lines are theoretical results. The results show that BER decreases significantly to 10^{-7} as SNR grows with a linear dynamic range. Moreover, no nonlinear distortion occurs at high signal power by using single-carrier M-QAM modulation for the 2 × 2 MIMO VLC system. We can see that BER of spatial diversity system has about 3-dB SNR gain compared to the spatial multiplexing, as the array gain in spatial diversity MIMO VLC improves the received signal strength. It is clear that the experimental measurements well match the theoretical results, which verify

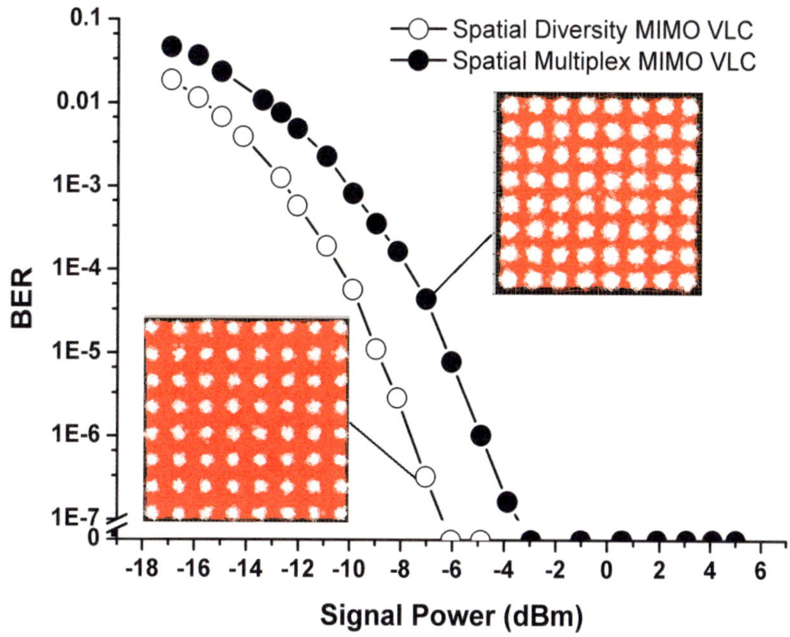

Figure 17. BER performance comparison for spatial diversity and spatial multiplex 64-QAM 2 × 2 MIMO VLC.

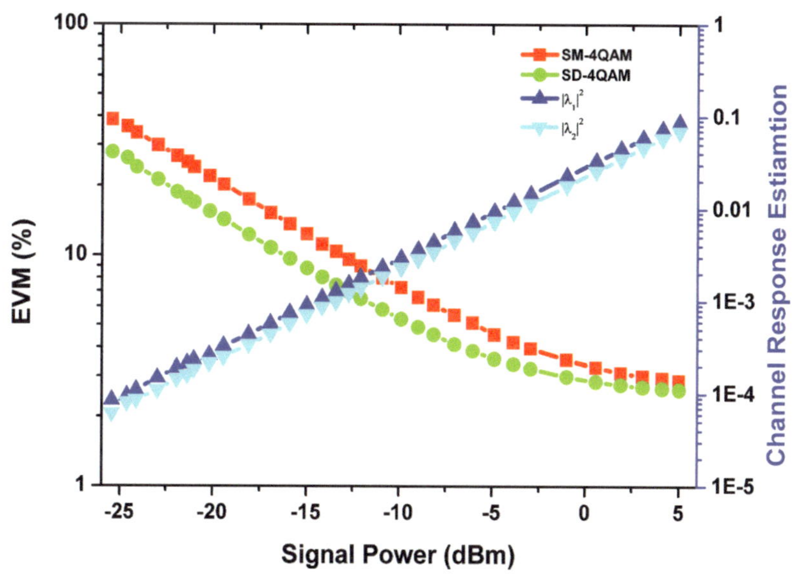

Figure 18. EVM and channel estimation against signal power for 2 × 2 MIMO VLC using spatial multiplexing (SM) and spatial diversity (SD).

accurate channel SNR estimation from the channel matrix in M-QAM MIMO VLC. Given the target BER threshold $BER_{tgt} = 10^{-3}$, we can establish adaptive MIMO mode criteria in terms of channel SNR and singular value of channel matrix.

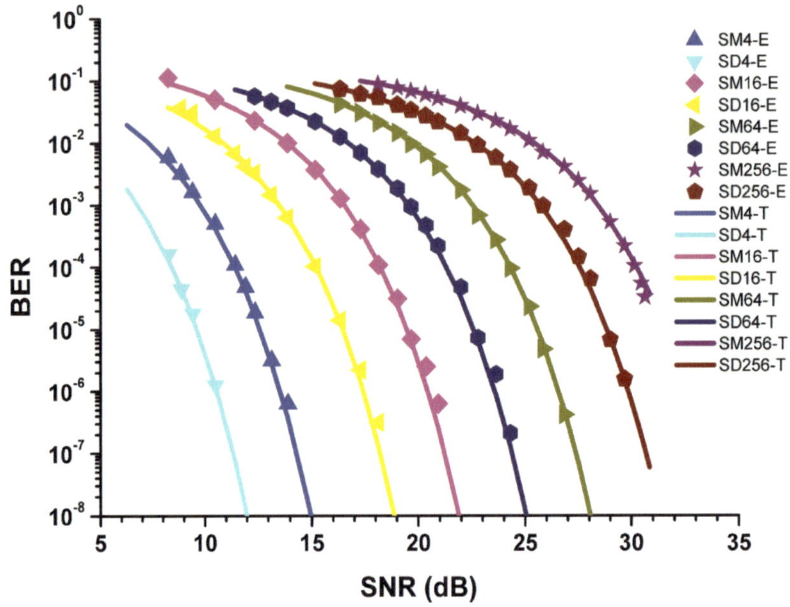

Figure 19. BER performance of M-QAM 2 × 2 MIMO VLC using spatial multiplexing and spatial diversity. Dots are experimental results and lines are theoretical results.

Figure 20 shows spectral efficiency performance of 4-16-64-256 QAM 2 × 2 MIMO VLC using spatial multiplexing and spatial diversity, where dots are experimental results and lines are theoretical results. Insets show the corresponding constellation diagrams. The results show spectral efficiencies of M-QAM MIMO VLC form step like shapes due to the rate discretization and BER integration. We can see that the spatial multiplexing system has about 3 dB gain in spectral efficiency compared with spatial diversity due to multiplexing gain of SM. These bandwidth efficiency enhancements are at the scarification of the degraded error performance. The experimental measurements well match the theoretical results, which verify channel SNR estimation and of spectral efficiency of M-QAM MIMO VLC. Thus, we can establish the adaptive MIMO mode criteria in terms of singular value of channel matrix to maximize the spectral efficiency.

Based on the established adaptive MIMO mode criteria in terms of channel estimation as shown in **Figure 21**, we demonstrate the real-time adaptive MIMO VLC system to maximize spectral efficiency under target BER threshold. BER performance and spectral efficiency against link distance for adaptive M-QAM MIMO VLC are depicted in **Figures 22** and **23**, respectively. The fixed 64-QAM MIMO VLC using spatial diversity and spatial multiplexing are introduced for comparison. Insets show the corresponding constellations of adaptive MIMO modes.

The initial mode is SM-64 (spatial multiplex 64-QAM). As we can see in the figures, as link distance increases from 0.6 to 1.4 m and 1.9 m, the adaptive modes change to SM-256, SM-64 and SM-16, receptively. Adaptive MIMO VLC increases the error free transmission distance to 2.2 m compared with 1.7 m for SM-64 and 1.9 m for SD-64. Furthermore, adaptive MIMO

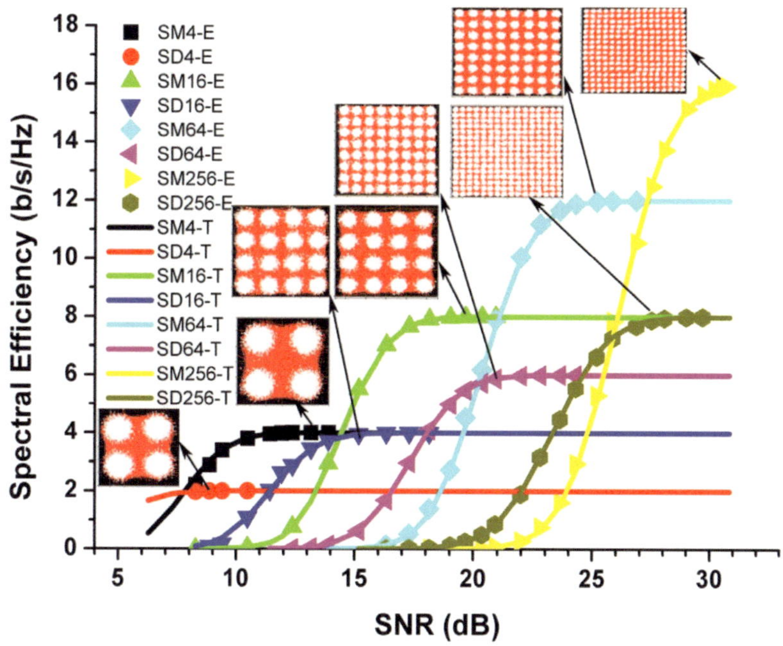

Figure 20. Spectral efficiency performance of M-QAM 2 × 2 MIMO VLC using spatial multiplexing and spatial diversity. Dots are experimental results and lines are theoretical results. Insets are the corresponding constellation diagrams.

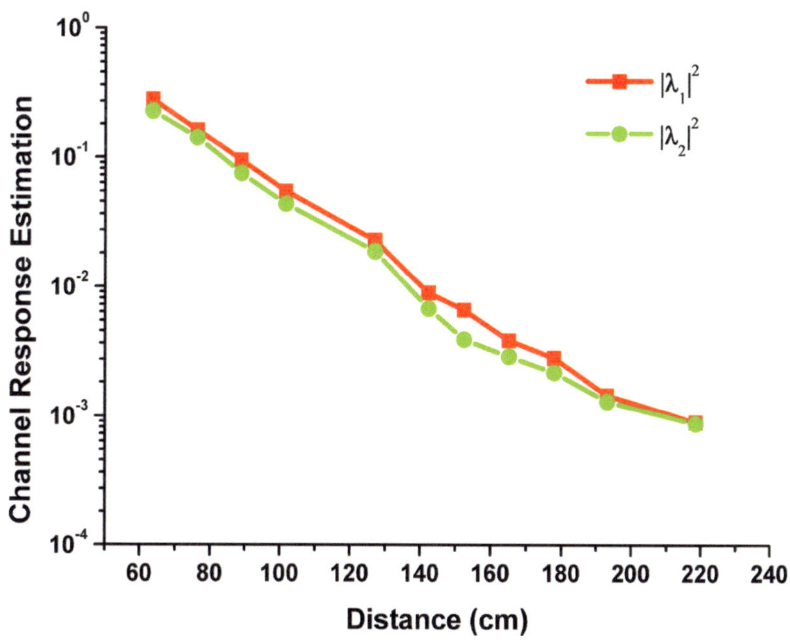

Figure 21. Channel estimation against link distance for 2 × 2 M-QAM MIMO VLC.

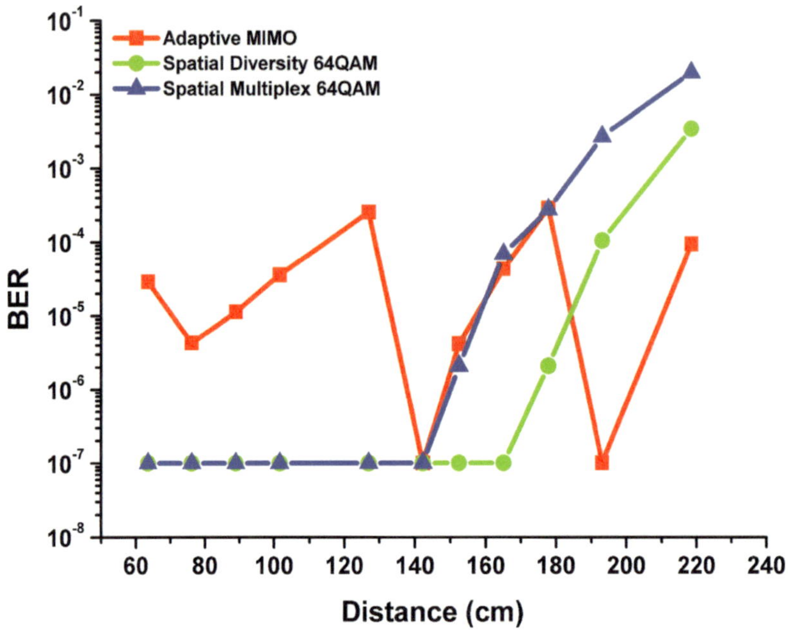

Figure 22. BER against link distance for adaptive M-QAM MIMO, spatial multiplexing 64-QAM, and spatial diversity 64-QAM VLC.

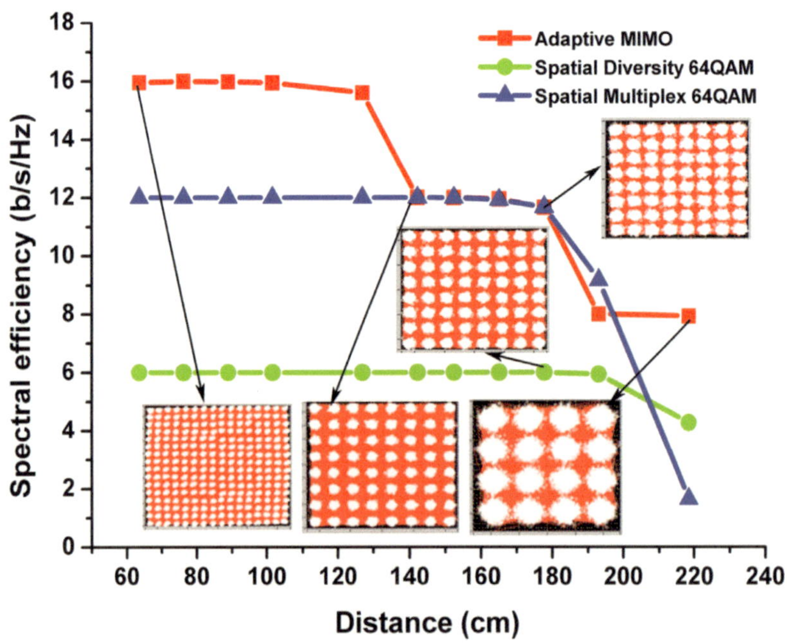

Figure 23. Spectral efficiency against link distance for adaptive M-QAM MIMO, spatial multiplexing 64-QAM, and spatial diversity 64-QAM VLC.

http://dx.doi.org/10.5772/intechopen.68919

VLC enhances the average spectral efficiency to 12 b/s/Hz over 2.2 m distance. Thus, it is necessary to explore the optimal combination of spatial multiplexing and spatial diversity for adaptive MIMO VLC to improve spectral efficiency and BER performance for indoor lighting environment.

4.3. Adaptive software-defined MIMO VLC after a mobile obstruction

Based on the established adaptive MIMO mode criteria in terms of channel estimation matrix, we demonstrate the real-time adaptive MIMO VLC system to maximize spectral efficiency under target BER threshold, after propagation beyond an obstruction. An obstacle with a diameter of 4.5 cm placed in the middle of transmission distance of 218 cm obstructs the line-of-sight light waves from two LEDs separated by 5 cm. We marked the central line between the two LEDs as a reference position at 0 cm. The obstacle was translated across the line-of-sight transmission direction from −65 to 65 cm, with dynamical obstruction range of 135 cm.

With the obstacle moving across LOS links, we measured the channel response of subchannels in **Figure 24**, and compared the error performance and spectral efficiency of adaptive M-QAM MIMO VLC with traditional MIMO VLC as shown in **Figures 25** and **26**. The fixed 64-QAM MIMO VLC using spatial diversity and spatial multiplexing are introduced for comparison. Insets show the corresponding constellations of adaptive MIMO modes. The initial mode is SM-64 (spatial multiplex 64-QAM). As we can see in the figures, as the obstacle moved across LOS from −65 to 65 cm, the adaptive MIMO modes change to SM-256, SM-64 and SD-256 adaptively with BER less than 10^{-3}. Results show that adaptive MIMO VLC can overcome these losses by the blockage, thus recover the clear constellation diagrams after propagation

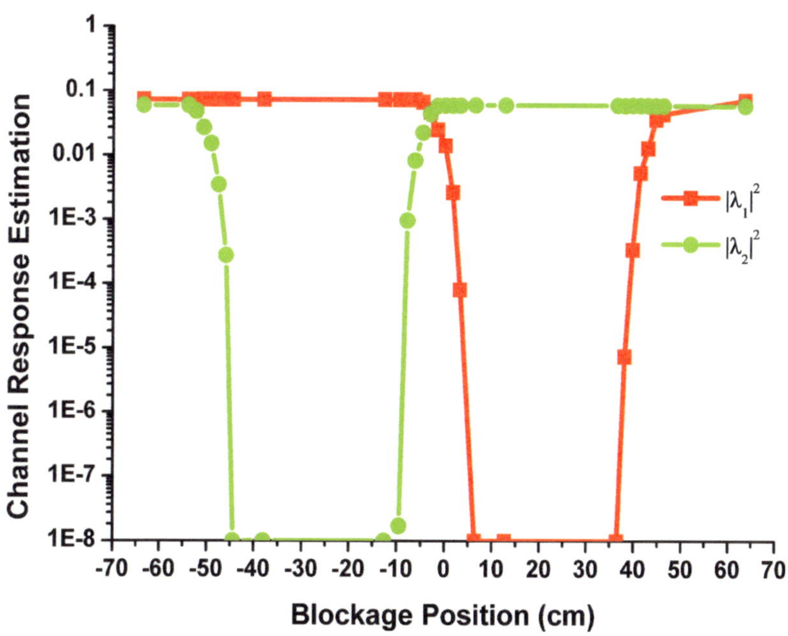

Figure 24. Channel estimation against blockage position for 2 × 2 M-QAM MIMO VLC.

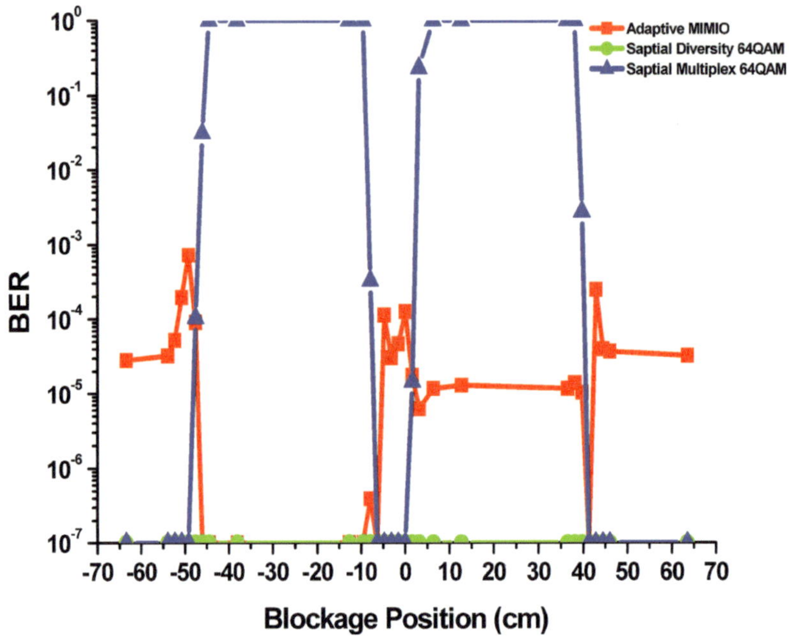

Figure 25. BER against blockage position for adaptive M-QAM MIMO VLC, spatial multiplexing 64-QAM, and spatial diversity 64-QAM VLC.

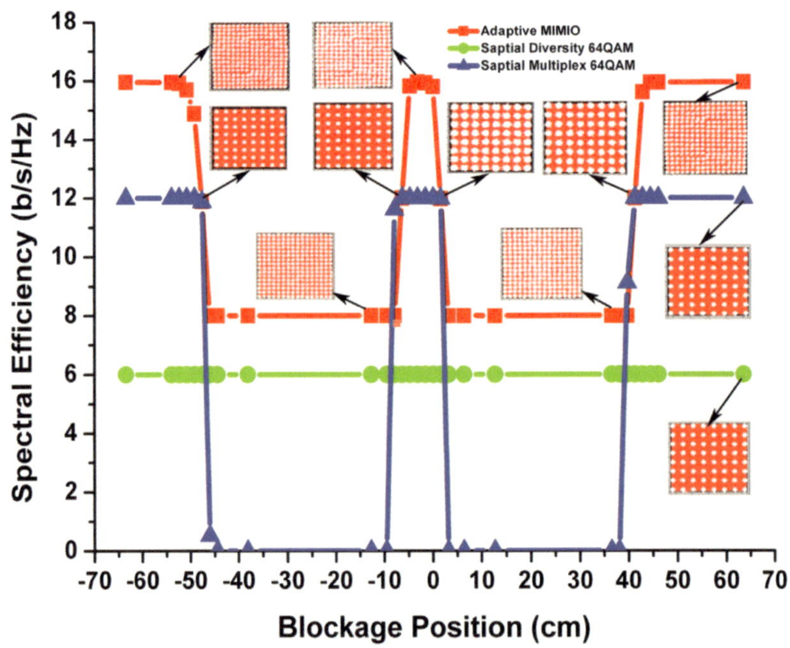

Figure 26. Spectral efficiency against blockage position for adaptive M-QAM MIMO VLC, spatial multiplexing 64-QAM, and spatial diversity 64-QAM VLC.

beyond the obstruction. Adaptive MIMO VLC enhanced the average error-free spectral efficiency to 12 b/s/Hz over 2-m indoor transmission with an obstruction, compared with 7.7 b/s/Hz for SM-64 and 6 b/s/Hz for SD-64. The adaptive MIMO VLC will enhance spectral efficiency and error performance in a real-time combined lighting and communication environment such as transmission power distribution, environmental blockage, and shadowing.

5. Conclusion

In this chapter, we experimentally demonstrate real-time software-defined multiple-input multiple-output (MIMO) visible light communication (VLC) system employing link adaptation of spatial multiplexing and spatial diversity. Real-time MIMO signal processing is implemented by using the field programmable gate array (FPGA)-based universal software radio peripheral (USRP) devices. Software-defined implantation of MIMO VLC can assist in enabling an adaptive and reconfigurable communication system without hardware changes. We propose the adaptive MIMO solution that both modulation schemas and MIMO schemas are dynamically adapted to the changing channel conditions for enhancing the error performance and spectral efficiency. The average error-free spectral efficiency of adaptive 2 × 2 MIMO VLC achieved 12 b/s/Hz over 2 m indoor dynamic transmission. The adaptive MIMO VLC will enhance performance in a real-time combined lighting and communication environment such as transmission power distribution, beam forming, communication range, environmental blockage, and shadowing.

Acknowledgements

This work has been funded by the National Science Foundation (NSF) ECCS directorate under projects Award #1201636 and Award #1160924.

Author details

Peng Deng

Address all correspondence to: dptyleo@gmail.com

Department of Electrical Engineering, The Pennsylvania State University, University Park, USA

References

[1] Kavehrad M. Sustainable energy-efficient wireless applications using light. IEEE Communications Magazine. 2010;**48**(12):66–73. DOI: 10.1109/Mcom.2010.5673074

[2] O'Brien D, Parry G, Stavrinou P. Optical hotspots speed up wireless communication. Nature Photonics. 2007;**1**(5):245–247. DOI: 10.1038/nphoton.2007.52

[3] Honglei L, Xiongbin C, Beiju H, Danying T, Hongda C. High bandwidth visible light communications based on a post-equalization circuit. Photonics Technology Letters, IEEE. 2014;**26**(2):119–122. DOI: 10.1109/LPT.2013.2290026

[4] Tsonev D, Chun H, Rajbhandari S, McKendry JJD, Videv S, Gu E, et al. A 3-Gb/s single-LED OFDM-based wireless VLC link using a gallium nitride μLED. Photonics Technology Letters, IEEE. 2014;**26**(7). DOI: 637–40. 10.1109/LPT.2013.2297621

[5] Wang Y, Tao L, Huang X, Shi J, Chi N. 8-Gb/s RGBY LED-Based WDM VLC System Employing High-Order CAP Modulation and Hybrid Post Equalizer. Ieee Photonics J. 2015;**7**(6):1–8. 10.1109/jphot.2015.2489927

[6] Neokosmidis I, Kamalakis T, Walewski JW, Inan B, Sphicopoulos T. Impact of Nonlinear LED Transfer Function on Discrete Multitone Modulation: Analytical Approach. Journal of Lightwave Technology. 2009;**27**(22):4970–78. DOI 10.1109/Jlt.2009.2028903

[7] Deng P, Kavehrad M, Kashani MA. Nonlinear modulation characteristics of white LEDs in visible light communications. In: Optical Fiber Communication Conference; 22–26 March 2015; Los Angeles, CA. 2015. p. W2A.64. DOI: 10.1364/OFC.2015.W2A.64

[8] Deng P, Kavehrad M, Liu Z, Zhou Z, Yuan X. Capacity of MIMO free space optical communications using multiple partially coherent beams propagation through non-Kolmogorov strong turbulence. Optics Express. 2013;**21**(13):15213–29.

[9] Azhar AH, Tran T, O'Brien D. A gigabit/s indoor wireless transmission using MIMO-OFDM visible-light communications. Photonics Technology Letters, IEEE. 2013;**25**(2):171–174. DOI: 10.1109/LPT.2012.2231857

[10] Burton A, Minh HL, Ghassemlooy Z, Bentley E, Botella C. Experimental Demonstration of 50 Mb/s Visible Light Communications using 4×4 MIMO. IEEE Photonic Tech L. 2014;**20** (9):945–8. DOI: 10.1109/LPT.2014.2310638

[11] Wang YQ, Chi N. Demonstration of high-speed 2 x 2 non-imaging MIMO Nyquist single carrier visible light communication with frequency domain equalization. Journal of Lightwave Technology. 2014;**32**(11):2087–2093. DOI: 10.1109/Jlt.2014.2320306

[12] Deng P, Yuan X, Kavehrad M, Zhao M, Zeng Y. Off-axis catadioptric fisheye wide field-of-view optical receiver for free space optical communications. Optical Engineering. 2012;**51** (6):063002.

[13] Te C, Lu L, Bo T, Zhong Z, Weiwei H. High-spatial-diversity imaging receiver using fisheye lens for indoor MIMO VLCs. Photonics Technology Letters, IEEE. 2014;**26**(22):2260–2263. DOI: 10.1109/LPT.2014.2354458

[14] Cui K, Chen G, Xu Z, Roberts RD. Line-of-sight visible light communication system design and demonstration. Communication Systems Networks and Digital Signal Processing

(CSNDSP), 2010 7th International Symposium on; July 21–23, 2010; Newcastle upon Tyne, UK. p. 621–5.

[15] Gu WJ, Aminikashani M, Deng P, Kavehrad M. Impact of multipath reflections on the performance of indoor visible light positioning systems. Journal of Lightwave Technology. 2016;**34**(10):2578–25787. DOI: 10.1109/Jlt.2016.2541659

[16] Deng P, Kavehrad M. Real-time software-defined single-carrier QAM mimo visible light communication system. 2016 Integrated Communications Navigation and Surveillance (ICNS); 19–21 April 2016; Herndon, VA. IEEE;2016. p. 5A3–5. DOI: 10.1109/ICNSURV.2016.7486354

[17] Deng P, Kavehrad M. Software Defined Adaptive MIMO Visible Light Communications after an Obstruction. Optical Fiber Communication Conference; 19-23 March 2017; Los Angeles, CA. OSA;2017. p. Th1E.5. DOI: 10.1364/OFC.2017.Th1E.5

[18] Rahaim M, Miravakili A, Ray S, Koomson V, Hella M, Little T. Software defined visible light communication. Wireless Innovation Forum Conference on Communications Technologies and Software Defined Radio (WInnComm SDR); 11-13 March, 2014. Schaumburg, Illinois.

[19] Deng P, Kavehrad M, Lou Y. MEMS-based beam-steerable free-space optical communication link for reconfigurable wireless data center. SPIE OPTO; January 28, 2017; San Francisco, CA; 2017. pp. 1012805–9.

[20] Deng P, Kavehrad M. Adaptive Real-Time Software Defined MIMO Visible Light Communications using Spatial Multiplexing and Spatial Diversity. Wireless for Space and Extreme Environments (WiSEE), 2016 IEEE International Conference on; 26–29 September 2016; Aachen, Germany. IEEE; 2016. p. 111–116. DOI: 10.1109/WiSEE.2016.7877314

[21] Zhang R, Wang J, Wang Z, Xu Z, Zhao C, Hanzo L. Visible light communications in heterogeneous networks: Paving the way for user-centric design. IEEE Wireless Communications. 2015;**22**(2):8–16. DOI: 10.1109/MWC.2015.7096279

[22] Elgala H, Mesleh R, Haas H. An LED model for intensity-modulated optical communication systems. IEEE Photonic Technology Letters. 2010;**22**(11):835–7. DOI: 10.1109/Lpt.2010.2046157

[23] Deng P, Kavehrad M. Effect of white LED DC-bias on modulation speed for visible light communications. arXiv preprint arXiv:161208477. 2016.

[24] Schubert EF, Gessmann T, Kim JK. Light emitting diode. 2nd ed. New York: Cambridge University Press; 2006. 422 p. DOI: 10.1017/CBO9780511790546

[25] Green RP, McKendry JJD, Massoubre D, Gu ED, Dawson MD, Kelly AE. Modulation bandwidth studies of recombination processes in blue and green InGaN quantum well micro-light-emitting diodes. Applied Physics Letters. 2013;**102**(9):1–4. Doi 10.1063/1.4794078

[26] Galler B, Drechsel P, Monnard R, Rode P, Stauss P, et al. Influence of indium content and temperature on Auger-like recombination in InGaN quantum wells grown on (111) silicon substrates. Applied Physics Letters. 2012;**101**(13):1–4. DOI: 10.1063/1.4754688

[27] Schubert EF, Hunt NEJ, Malik RJ, Micovic M, Miller DL. Temperature and modulation characteristics of resonant-cavity light-emitting diodes. Journal of Lightwave Technology. 1996;**14**(7):1721–1729. DOI: 10.1109/50.507950

[28] Komine T, Nakagawa M. Fundamental analysis for visible-light communication system using LED lights. IEEE Trans/actions on Consumer Electronics. 2004;**50**(1):100–107.

[29] Oestges C, Clerckx B. MIMO wireless communications: From real-world propagation to space-time code desig. Orlando: Academic Press; 2010. 480 p.

[30] Zhou ZD, Vucetic B, Dohler M, Li YH. MIMO systems with adaptive modulation. IEEE on Transactions on Vehicular Technology. 2005;**54**(5):1828–1842. DOI: 10.1109/tvt.2005.853886

Chapter 3

Transceiver Design for MIMO DCO-OFDM in Visible Light Communication

Jian Dang, Mengting Wu, Liang Wu and Zaichen Zhang

Additional information is available at the end of the chapter

http://dx.doi.org/10.5772/intechopen.68887

Abstract

Direct current-biased optical-orthogonal frequency-division multiplexing (DCO-OFDM) is a simple yet spectrally efficient multicarrier modulation scheme for visible light communication (VLC). But in multiple-input multiple-output (MIMO) scenario, which is more practical for VLC due to the LED deployment, the research on DCO-OFDM is still limited and calls for in-depth investigation. In this chapter, we first study the basic modulation scheme of DCO-OFDM, including the design of conventional receiver without considering the clipping noise. Secondly, we present a novel receiver for combating clipping distortion in the DCO-OFDM system, which can reconstruct the clipping noise and subtract it from the received signal. Thirdly, we generalize the results to MIMO scenario and investigate the preliminary transceiver design, which is based on the minimum mean-square error (MMSE) criteria. Based on this, we propose a precoding algorithm to further enhance the performance. Finally, the symbol error rate performance is compared through computer simulations to give the reader a whole picture of the performance of MIMO VLC system.

Keywords: clipping noise, DCO-OFDM, iterative reconstruction, transceiver design, DCO-OFDM

1. Introduction

As an important complementary technique to the fifth generation (5G) wireless communication system, visible light communication (VLC) is currently enjoying a promising development in the research area of communication [1]. VLC transmits information with the visible light emitted by the light emitting diodes (LEDs), which are generally deployed to provide indoor illumination simultaneously. In VLC, simple low-cost intensity modulation and direct detection (IM/DD) techniques are employed to send information, which implies that the phase cannot be used to convey any information. Therefore, pulsed modulation schemes such as pulse position modulation (PPM), pulse width modulation (PWM) or on-off keying (OOK) is employed in practical VLC systems [2].

In order to combat the inter-symbol interference caused by the indoor optical wireless channel, orthogonal frequency-division multiplexing (OFDM) has been considered for VLC due to its inherent robustness to multipath effect [3]. In VLC, the transmitted electrical signal is used to modulate the light intensity of the LED. For this purpose, the transmitted electrical signal must be real and non-negative. Therefore, the traditional OFDM needs to be modified to obtain the real-valued signal, which can be produced by imposing Hermitian symmetry symbols on the frequency domain subcarriers and then conducting inverse discrete Fourier transform (IDFT) operation on these frequency domain subcarriers. In general, there are two approaches to obtain non-negative signal from the above real-valued signal. One approach is asymmetrically clipped optical OFDM (ACO-OFDM) which is proposed in Ref. [4]. The other approach is the direct current (DC)-biased optical (DCO-OFDM), which is more spectrally efficient than ACO-OFDM [5]. The differences between ACO-OFDM and DCO-OFDM are explained in more detail in Refs. [5, 6].

One major problem of OFDM modulation is its high peak-to-average power ratio (PAPR), which results from the summation over a large number of subcarriers. Therefore, the time domain DCO-OFDM signal often has to be double-sided clipped to fit the dynamic linear range of the transmitter, which inevitably introduces clipping distortion [7]. Although some papers have analysed the double-sided clipped DCO-OFDM, to our knowledge, few of them focus on improving the performance of the receiver [8]. In Ref. [9], the authors only analysed the performance of the double-sided clipped DCO-OFDM in terms of the achievable data rate and error vector magnitude. In Ref. [8], an optimum DC bias method has been proposed for the double-sided clipped DCO-OFDM in order to decrease the clipping effects of the transmitter. In Refs. [3–9], only the conventional receiver is utilized in DCO-OFDM. The conventional receiver for DCO-OFDM usually does not care for the clipping noise and only simply subtracts the DC bias from the received signal in the process of demodulation. Not surprisingly, it can only provide suboptimal performance, especially in heavily clipped scenarios. In this chapter, we are interested in designing the high-performance receiver for DCO-OFDM. First, we analyse the structure of the transmitted signal. Then, the clipping noise can be modelled as a certain signal format that is based on the lower and upper clipping bounds. At last, we propose a new receiver for DCO-OFDM, which reconstructs the clipping noise and decodes data in an iterative manner by exploiting the signal structure of the clipping noise. The complexity of the new receiver is comparable with the conventional one. Moreover, the convergence rate of the new receiver is very fast, about 2–5 iterations. Therefore, the proposed receiver can provide a realistic implementation in the DCO-OFDM system. Simulation results demonstrate the large performance gains compared to the conventional receiver.

On the other aspect, the low-cost white LED has limited modulation bandwidth, which prohibits high transmission rate of VLC, and optical multiple-input multiple-output (MIMO) can help achieve high data rate by utilizing spatial diversity [10]. Therefore, OFDM can also be deployed together with MIMO in VLC systems to further enhance the bandwidth utilization efficiency and boost the data rates [11, 12]. Some experimental results have demonstrated that high data rates up to gigabit/s can be achieved in the MIMO VLC system [12]. In this chapter, we first describe a MIMO VLC system, which can effectively support the flickering/dimming control and other lighting requirements [13]. Secondly, in order to drive an LED transmitter, we illustrate how the dynamic range constraint and dimming control impact the design of MIMO transceiver [14]. Finally, an advanced method on MIMO transceiver design is proposed and studied by theoretical analysis and simulation. It turns out that the bit error rate (BER) performance of the MIMO VLC system could be improved by the advanced joint optimization method.

For convenience, we summarize the mathematical notations here. Bold and lowercase symbols (e.g. $\boldsymbol{a}$) denote column vectors. Specifically, **1** denotes a column vector whose elements are 1. Bold and uppercase symbols (e.g. $\boldsymbol{H}$) denote matrices. Specifically, $\boldsymbol{W}_N$ denotes the N × N DFT matrix. The superscripts $(\cdot)^*$, $(\cdot)^{\mathrm{T}}$, $(\cdot)^{\mathrm{H}}$ and $(\cdot)^{-1}$ denote the complex conjugate, transpose, conjugate transpose and inverse of a matrix/vector, respectively. sign(a) returns the signs of a (here 0 is replaced by 1), $E\{\cdot\}$ denotes the statistical expectation, diag(·) is a diagonal matrix whose entries on the main diagonal are given by a matrix, Tr(·) is the trace of a matrix, $\|\cdot\|$ denotes the Frobenius norm, min(·) denotes an element-wise minimum operator and abs(·) denotes an element-wise absolute operator.

2. Conventional receiver for DCO-OFDM

2.1. System model of DCO-OFDM

The DCO-OFDM system model considered in this chapter is shown in **Figure 1**. At the transmitter, the source bits $\boldsymbol{b}$ (coded or uncoded) are first modulated using M-ary square quadrature amplitude modulation (QAM) to get the complex signal $\boldsymbol{X}$. In order to get the real output of time domain OFDM signal after IDFT operation, the input frequency domain vector $\tilde{\boldsymbol{X}}$ of the IDFT module must be symmetrically conjugated, which requires Hermitian symmetric expansion of $\boldsymbol{X}$. The specific expansion is given as follows:

$$\boldsymbol{X} = \left[X(0), X(1), \ldots, X\left(\frac{N}{2}-1\right)\right]^T, X(0) = 0, \tag{1}$$

$$\tilde{\boldsymbol{X}} = [\tilde{X}(0), \tilde{X}(1), \ldots, \tilde{X}(N-1)]^T, \tag{2}$$

$$\tilde{X}(k) = \begin{cases} X(K), & 0 \le k \le \dfrac{N}{2} - 1 \\ 0, & k = \dfrac{N}{2} \\ X^*(n-k), & \dfrac{N}{2} + 1 \le k \le N-1 \end{cases}. \tag{3}$$

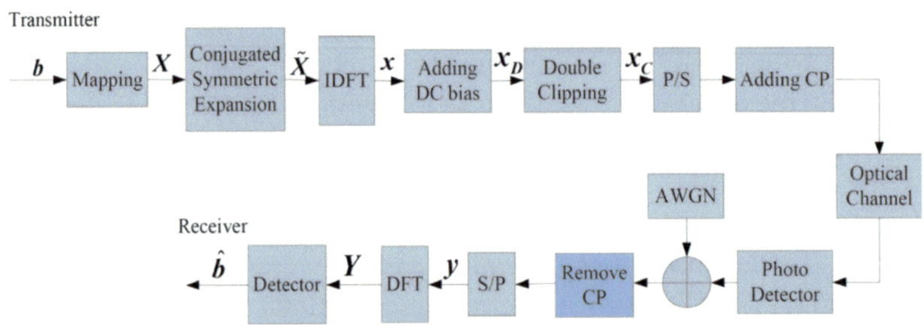

Figure 1. A DCO-OFDM system model.

Here, $\tilde{X}(0)$ and $\tilde{X}(N/2)$ do not carry information and are generally set to zeros. Thus, the number of unique data carrying subcarriers present is $\frac{N}{2}-1$. Then, the discrete time-domain real-valued OFDM signal $\boldsymbol{x}$ is generated by applying IDFT operation to $\tilde{\boldsymbol{X}}$ as

$$x(n) = \frac{1}{\sqrt{N}}\sum_{k=0}^{N-1} \tilde{X}(k)\exp\left(\frac{j2\pi kn}{N}\right), n = 0, 1, \ldots, N-1, \tag{4}$$

where $j=\sqrt{-1}$ and N is the size of IDFT. When N is not very small, according to the central limit theorem, it is reasonable to model $\boldsymbol{x}$ as an independent identical distributed (i.i.d) Gaussian vector with zero mean. In order to drive the LED and fit the dynamic linear range of the transmitter simultaneously, the signal should be added with a proper DC bias, which leads to $\boldsymbol{x}_D$ as

$$\boldsymbol{x}_D = \boldsymbol{x} + \mathbf{1}\cdot D, \tag{5}$$

where D is the DC bias. In DCO-OFDM, double-sided clipping should be applied to $\boldsymbol{x}_D$ to fit the dynamic linear range of the LED. Then, $\boldsymbol{x}_D$ needs to be clipped at both lower and upper bounds to get $\boldsymbol{x}_C$, which can be described as

$$x_C(n) = \begin{cases} C_L, & x_D(n) < C_L \\ x_D(n), & C_L \leq x_D(n) \leq C_U \\ C_U, & x_D(n) > C_U \end{cases} \tag{6}$$

where C_L and C_U are the lower and upper clipping bounds, respectively, which need to satisfy the constraint of $C_L \leq x_D(n) < C_U$. We note that Eq. (6) can be rewritten as

$$x_C(n) = x(n)f_1\left(x(n)\right) + D\cdot f_1\left(x(n)\right) + C_L\cdot f_2\left(x(n)\right) + C_U\cdot f_3\left(x(n)\right) \tag{7}$$

by defining

$$f_1\left(x(n)\right) = \frac{sign\left(x(n)+D-C_L\right)+1}{2}\cdot\frac{sign\left(C_U-x(n)-D\right)+1}{2}, \tag{8}$$

$$f_2\left(x(n)\right) = \frac{sign\left(C_L-x(n)-D\right)+1}{2}, \tag{9}$$

http://dx.doi.org/10.5772/intechopen.68887

$$f_3\left(x(n)\right) = \frac{sign\left(x(n) + D - C_U\right) + 1}{2}. \tag{10}$$

In vector form:

$$\boldsymbol{x}_C = \boldsymbol{F}_1\boldsymbol{x} + \boldsymbol{f}_1 \cdot D + \boldsymbol{f}_2 \cdot C_L + \boldsymbol{f}_3 \cdot C_U, \tag{11}$$

$$\boldsymbol{F}_1 = diag(\boldsymbol{f}_1) = diag\left(f_1\left(x(0)\right), f_1\left(x(1)\right), \ldots, f_1\left(x(N-1)\right)\right). \tag{12}$$

At the transmitter, $\boldsymbol{x}_C$ is appended with the cyclic prefix (CP) to avoid inter-symbol interference and then drives the LED.

The optical signal from the LED spreads through the optical channel and then arrives at the receiver, where a photodetector (PD) can retrieve the electrical signal on the basis of the intensity of the received optical signal. After removing CP, the received signal in time domain and frequency domain after DFT is denoted as $\boldsymbol{y}$ and $\boldsymbol{Y}$, respectively. In VLC, the equivalent electrical domain channel between $\boldsymbol{x}_C$ and $\boldsymbol{y}$ is denoted by $\boldsymbol{h}$, which may be a direct path and/or reflected paths. All the components of $\boldsymbol{h}$ are real and non-negative. The frequency domain channel of $\boldsymbol{h}$ is denoted as $\boldsymbol{H}$, and we define $\boldsymbol{H} = diag(\boldsymbol{H})$, which is based on the principles of OFDM system and the singular value decomposition (SVD) of circle matric. The received signal in the frequency domain can be expressed as

$$\boldsymbol{Y} = \boldsymbol{H}\boldsymbol{X}_C + \boldsymbol{Z} \tag{13}$$

where $\boldsymbol{X}_C = \boldsymbol{W}_N\, \boldsymbol{x}_C$ and $\boldsymbol{Z}$ is the additive white Gaussian noise in the frequency domain and each of its sample has zero mean and variance σ^2. To facilitate the derivation of the new receiver in subsection III, we rewrite Eq. (13) by combining Eqs. (7)–(12) as

$$\begin{aligned} \boldsymbol{Y} &= \boldsymbol{H}\boldsymbol{W}_N(\boldsymbol{F}_1 x + \boldsymbol{f}_1 \cdot D + \boldsymbol{f}_2 \cdot C_L + \boldsymbol{f}_3 \cdot C_U) + \boldsymbol{Z} \\ &= \boldsymbol{H}\boldsymbol{W}_N\boldsymbol{F}_1\boldsymbol{W}_N^H\boldsymbol{X} + \boldsymbol{H}\boldsymbol{W}_N\boldsymbol{f}_1 \cdot D + \boldsymbol{H}\boldsymbol{W}_N\boldsymbol{f}_2 \cdot C_L + \boldsymbol{H}\boldsymbol{W}_N\boldsymbol{f}_3 \cdot C_U + \boldsymbol{Z}. \end{aligned} \tag{14}$$

It is worth noting that DC bias D should be carefully designed to reduce distorting the signal $\boldsymbol{x}$ as little as possible [8], which is beyond the scope of this chapter.

2.2. Conventional receiver design of DCO-OFDM

For the conventional receiver, it does not care for the clipping noise, that is, it models $\boldsymbol{x}_C$ as follows:

$$\boldsymbol{x}_C \approx \boldsymbol{x}_D = \boldsymbol{x} + \boldsymbol{1} \cdot D. \tag{15}$$

Thus, the received signal in the frequency domain is

$$\boldsymbol{Y} \approx \boldsymbol{H}\boldsymbol{W}_N(\boldsymbol{x} + \boldsymbol{1} \cdot D) + \boldsymbol{Z} = \boldsymbol{H}\tilde{\boldsymbol{X}} + \boldsymbol{H}\boldsymbol{W}_N\boldsymbol{1} \cdot D + \boldsymbol{Z}. \tag{16}$$

At the receiver, the DC bias is first removed by

$$Y_1 = Y - HW_N\mathbf{1} \cdot D \approx H\tilde{X} + Z. \tag{17}$$

Then, based on Eq. (17), the receiver can decode the data. This is a common estimation problem in wireless communications and there are many available estimation methods. Without loss of generality, in this chapter, we select the linear minimum mean-square error (LMMSE) estimator because of its good complexity-performance trade-off. The estimate of $\tilde{X}$ is given by

$$\tilde{X}_{\text{conv}} = (H^H H + \sigma^2 I_N)^{-1} H^H Y_1 \tag{18}$$

3. Advanced receiver for DCO-OFDM

3.1. Advanced receiver design of DCO-OFDM

While the conventional receiver is simple to implement, it does not suppress the clipping noise in the process of demodulation, which inevitably degrades the system performance severely, especially in a heavily clipped case. Here, we propose a novel receiver, which iteratively reconstructs the clipping noise using the structure of the clipping noise and subtracts it from the received signal. Then, one can get the signal with less clipping noise. Subsequently, we conduct the LMMSE on the signal with less chipping noise to obtain the more accurately detected source bits $\hat{b}$. Based on Eq. (14), the clipping noise is given as

$$Z_C = HW_N f_1 \cdot D + HW_N f_2 \cdot C_L + HW_N f_3 \cdot C_U. \tag{19}$$

Here in Eq. (19), reconstructing Z_C is dependent on the data $\tilde{X}$ since f_1, f_2 and f_3 are functions of x. Therefore, we propose the following iterative algorithm to decode the data.

Algorithm 1: Proposed iterative for DCO-OFDM

Step 1. Initialization:

Using the conventional receiver that is based on Eq. (18), obtain the estimation of $\tilde{X}$, denoted by $\tilde{X}_{\text{conv}}$. Then, reconstruct the time domain signal $\hat{x}$ which is an approximation of x by performing IDFT operation on $\tilde{X}_{\text{conv}}$.

Step 2. Iteration:

1. Based on Eqs. (8)–(10) and $\hat{x}$, obtain the estimated values of f_1,f_2 and f_3;
2. Based on Eq. (19), construct the estimated clipping noise $\tilde{Z}_C$, and then subtract $\tilde{Z}_C$ from the received frequency domain signal Y to get the estimate of

$$Y_{sci} = Y - \tilde{Z}_C \approx HW_N F_1 W_N^H X + Z.$$

3. If F_1 is full rank, go to (4) or otherwise go to (5).

4. Perform LMMSE on $\boldsymbol{Y}_{sci}$ to obtain $\tilde{\boldsymbol{X}}_{prop}$, which is an updated estimation of $\tilde{\boldsymbol{X}}$. Then, construct $\hat{\boldsymbol{x}}$ based on $\tilde{\boldsymbol{X}}_{prop}$ by performing IDFT operation;

5. Based on $\boldsymbol{W}_N^H\boldsymbol{H}^{-1}\boldsymbol{Y}_{sci} \approx \boldsymbol{F}_1\boldsymbol{x} + \boldsymbol{W}^H\boldsymbol{H}^{-1}\boldsymbol{Z}$, or $\boldsymbol{y}_{sci} \approx \boldsymbol{F}_1\boldsymbol{x} + z_{sic}$, if $\boldsymbol{F}_1(n, n) = 0$, replace $y_{sic}(n)$ with the former estimation of $x(n)$. Through the updated $\boldsymbol{y}_{sci}$, obtain $\tilde{\boldsymbol{X}}_{prop}$, and then construct $\hat{\boldsymbol{x}}$.;

6. If the algorithm has reached a predefined number of iterations, go to (3) or otherwise go to (1).

Step 3. Demapping:

Extract $\tilde{\mathbf{X}}_{prop}$ with indices 1, 2, …, $N/2 - 1$ and then perform demapping to get the source bit estimates $\hat{\boldsymbol{b}}$.

In the following, we address some issues in the process of implementing the proposed receiver that one should pay attention to.

1. $f_1(\cdot)$ is 0,which corresponds to the negative clipping effects. Thus, $\boldsymbol{F}_1$ cannot be inverted. Therefore, if we suppose that $\boldsymbol{H}\boldsymbol{W}_N\boldsymbol{F}_1\boldsymbol{W}_N^H$ could be inverted, using $\boldsymbol{Y}_{sci} = \boldsymbol{Y} - \tilde{\boldsymbol{Z}}_C \approx \boldsymbol{H}\boldsymbol{W}_N\boldsymbol{F}_1\boldsymbol{W}_N^H\boldsymbol{X} + \boldsymbol{Z}$ with LMMSE will result in bad performance.

2. To get rid of the inversion of $\boldsymbol{F}_1$, we use the following methods: $\boldsymbol{W}_N^H\boldsymbol{H}^{-1}\boldsymbol{Y}_{sci} \approx \boldsymbol{F}_1\boldsymbol{x} + \boldsymbol{W}^H\boldsymbol{H}^{-1}\boldsymbol{Z}$ or $\boldsymbol{y}_{sci} \approx \boldsymbol{F}_1\boldsymbol{x} + z_{sic}$. If $\boldsymbol{F}_1(n, n) = 0$; then no information could be obtained for $x(n)$ from $y_{sic}(n)$; thus, we use former estimation of $x(n)$ If $\boldsymbol{F}_1(n, n) = 1$, then $x(n)$ could be reestimated by $y_{sic}(n)$. After all, $x(n)$ is obtained and $\tilde{X}$ could be obtained.

3.2. Numerical results of DCO-OFDM

In this section, we show numerical results of the average uncoded BER performance for the proposed receiver under different double-sided clipping bounds and different modulation sizes of QAM and compare it with the conventional receiver. The channels are generated using the method in Ref. [2] with the following parameters: an empty room of size 8 m × 6 m × 4 m, the reflection coefficients for the ceiling, the wall and the floor which are 0.8, 0.8 and 0.3, respectively and the LED is attached 0.1 m below the ceiling and the photodetector is 1 m above the floor with an 80 degree of field of view (FOV). The LED pointing straight downward and upward can generate line-of-sight (LOS) and non-line-of-sight (NLOS) channels, respectively. The channels are normalized to have the power 1. The transmit power of x is normalized to 1, that is $\sigma_x = 1$. The upper and lower clipping bounds are set to be linearly proportional to σ_x. The number of subcarriers is $N = 64$. The DC bias is set to be $3\sigma_x$.

Figures 2 and **3** present the BER performance with the different double-sided clipping bounds and the fixed modulation size $M = 16$, in LOS and NLOS channels, respectively. **Figure 4** shows the BER performance with the different modulation sizes and the fixed double-sided clipping bound ($C_L = 0$, $C_U = 6\sigma_x$) in LOS channels. **Figure 5** presents the BER performance with the different modulation sizes and the fixed double-sided clipping bound ($C_L = 0$, $C_U = 5\sigma_x$) in NLOS channels. Through simulations, it is observed that the proposed receiver converges after only 2 iterations for most times. Therefore, 2 iterations are involved for the proposed receiver.

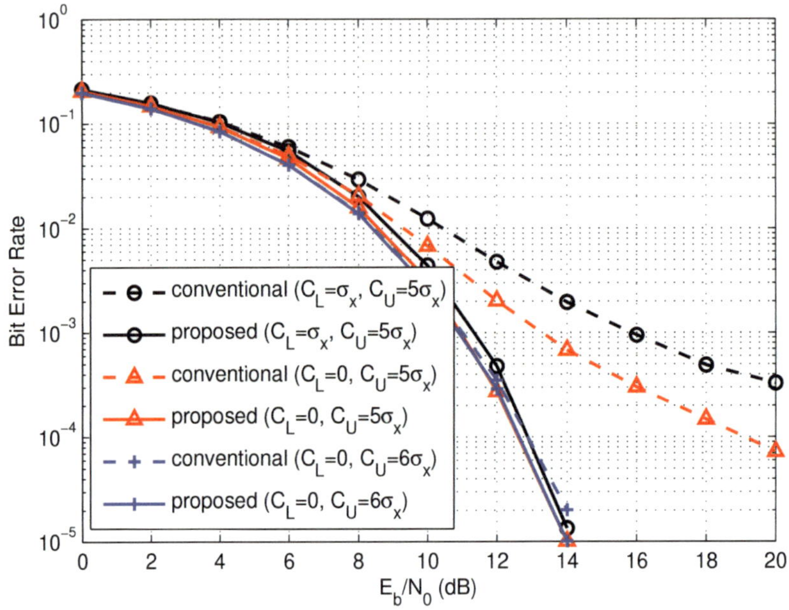

Figure 2. BER comparison between the proposed receiver and the conventional receiver with different lower and upper clipping bounds and fixed modulation size $M = 16$ in LOS channels.

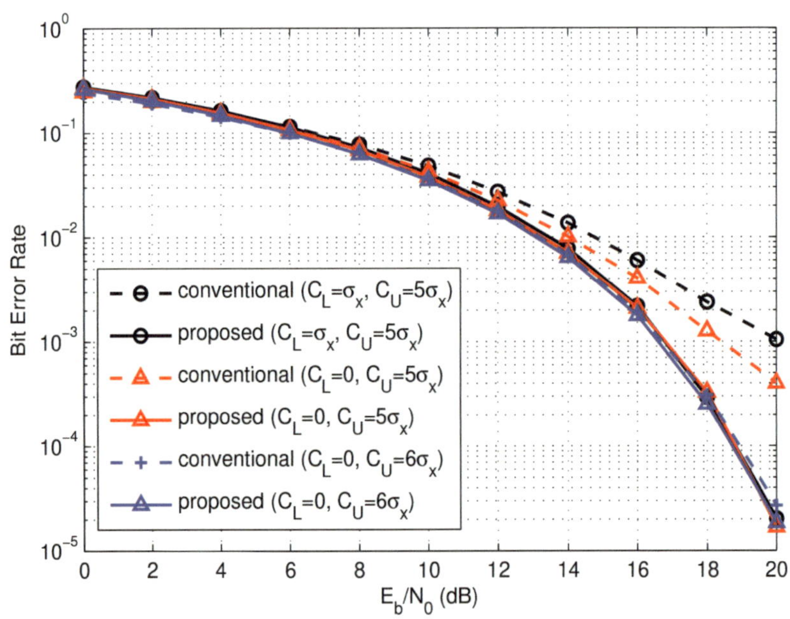

Figure 3. BER comparison between the proposed receiver and the conventional receiver with different lower and upper clipping bounds and fixed modulation size $M = 16$ in NLOS channels.

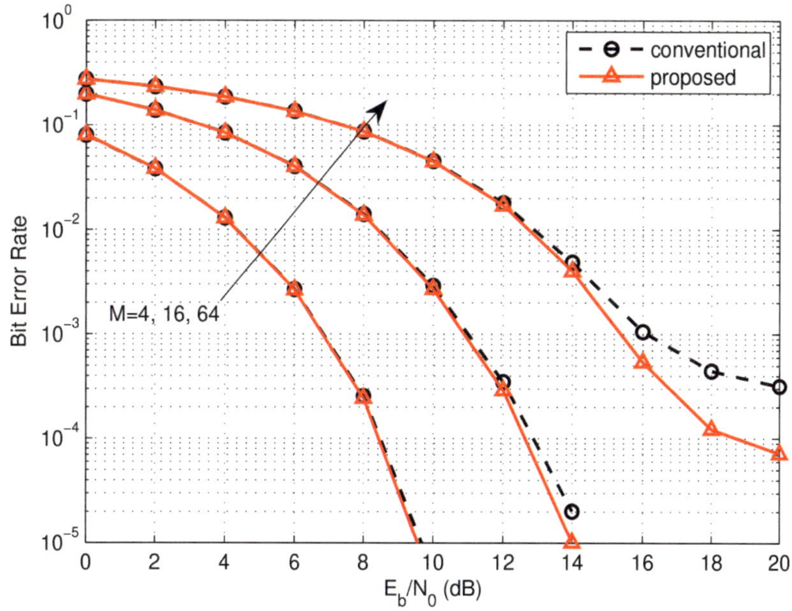

Figure 4. BER comparison between the proposed receiver and the conventional receiver with different modulation sizes and the fixed lower and upper clipping bounds ($C_L = 0$, $C_U = 6\sigma_x$) in LOS channels.

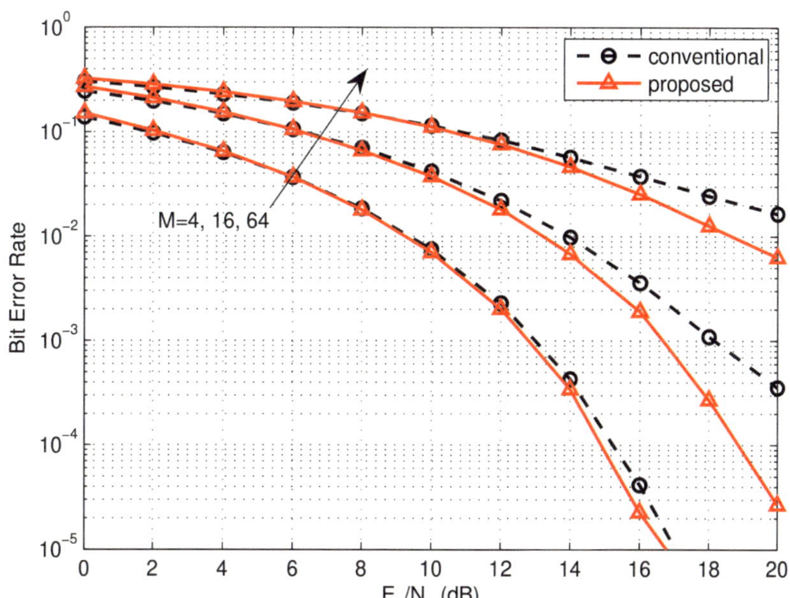

Figure 5. BER comparison between the proposed receiver and the conventional receiver with different modulation sizes and the fixed lower and upper clipping bounds ($C_L = 0$, $C_U = 5\sigma_x$) in NLOS channels.

We can obtain several facts from **Figures 2–5**. First, the proposed receiver outperforms the conventional one in any case, that is, in both LOS and NLOS and with different double-sided clipping bounds and different modulation sizes. Second, the signal noise ratio (SNR) gain over the conventional receiver could be much more than 3 dB, especially when decreasing the clipping range. Although the SNR gain over the conventional receiver could fade away when increasing the clipping range, the larger clipping range requires the larger dynamic linear range of the LED, which is difficult to satisfy in practice. Finally, the proposed receiver can give better performance when higher order QAM formats are employed. Therefore, the proposed receiver can provide better performance than the conventional receiver.

4. Transceiver design for MIMO DCO-OFDM

In the field of MIMO VLC system, the joint design of precoder and equalizer still need deep research. In previous works, a simple way to design the transceiver in MIMO VLC systems was to apply the channel matrix inversion at the receiver side [12]. If the channel matrix is rank-deficient or not squared, pseudo-inversion operation is used instead. However, these methods might result in noise amplification if the values of elements in the channel matrix were low [15]. In Refs. [13, 14], a MIMO VLC system was proposed, which can effectively support the flickering/dimming control and other lighting requirements. Moreover, in such a system, the precoder and equalizer can be adaptively optimized according to different input signals and various illumination levels. Based on Ref. [14], we propose an advanced transceiver design on MIMO VLC systems, which will be detailed in this section.

4.1. System model of MIMO DCO-OFDM

A MIMO DCO-OFDM system for optical wireless communication is surveyed and simulated in Ref. [16], and from this research, it is known that an MIMO DCO-OFDM system can be equivalently split into N parallel MIMO subsystems with single-tap channels in the frequency domain, where N is the number of subcarriers. Therefore, for simplicity and without loss of generality, we only investigate a specific subcarrier of the MIMO DCO-OFDM system which can be modelled as a single-path MIMO subsystem. The result can be readily extended to the whole MIMO DCO-OFDM system. **Figure 6** shows the simplified MIMO subsystem with N_t LEDs at the transmitter and N_r photodetectors at the receiver. The optical MIMO channel is

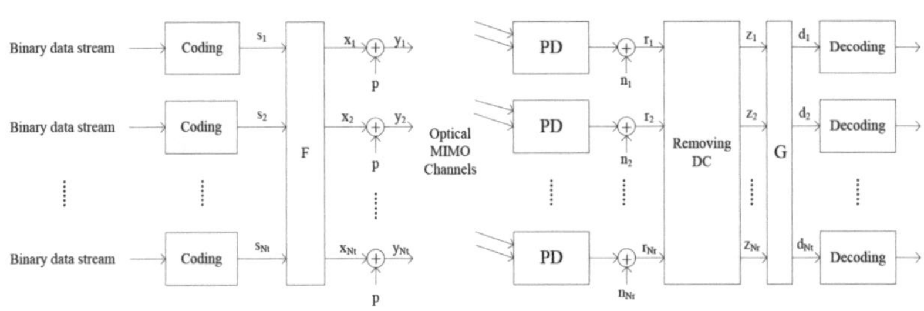

Figure 6. The simplified MIMO DCO-OFDM system model.

modelled as an $N_t \times N_r$ matrix, whose element h_{mn} is real-valued and represents the channel gain between the nth LED and the mth PD.

Here, we assume that the channel response in each transmission period is flat, and the multi-level pulse amplitude modulation (PAM) is a practical example for this section. Moreover, PPM or PWM is not considered here because different pulse positions and various pulse widths (caused by dimming control) from different LEDs can make MIMO detection more complicated than the case of PAM. Since VLC employs IM/DD technology, the signals must be real-valued. Thus, all vectors and matrices are real-valued in this chapter.

4.1.1. Transmitter side

From **Figure 6**, we know that N_t binary data streams are firstly modulated into the source data vector $\boldsymbol{s}(\tau) = [s_1(\tau), \ldots, s_{N_t}(\tau)]^T$ For convenience, we ignore the time index τ in the rest of this chapter. Taking into account the actual situation, we assume that the source data s_k is zero mean and bounded as

$$-b_k \le s_k \le b_k. \tag{20}$$

Because of the characteristics of the optical channel, the source data streams cannot drive the LEDs directly. To meet dimming control and the dynamic-range constraints of LEDs, the source data vector will be multiplied by an $N_t \times N_r$ precoder matrix $\boldsymbol{F}$ and then added by the DC biasing $\boldsymbol{p} = p \times 1$, which is shown in **Figure 6**. Such a MIMO VLC transmitter structure was firstly introduced in Ref. [13] but only the real-valued LED transmission signals were considered.

The transmitted signal can be expressed as $\boldsymbol{y} = \boldsymbol{Fs} + \boldsymbol{p}$, where we assume that the LED electrical-to-optical conversion is linearized [17] and the overall gain is chosen as 1 for convenience without loss of generality. For each LED, we assume that $[l, u]$ $(0 < l)$ is the dynamic-range constraint, and in order to meet the dimming control, we should have

$$0 < l \le y_i = \sum_{k=1}^{N_t} f_{ik} s_k + p \le u. \tag{21}$$

Combining with Eq. (20), we know that the signal after precoder satisfies

$$-\sum_{k=1}^{N_t} |f_{ik}| b_k \le \sum_{k=1}^{N_t} f_{ik} s_k \le \sum_{k=1}^{N_t} |f_{ik}| b_k. \tag{22}$$

So, we need

$$\sum_{k=1}^{N_t} |f_{ik}| b_k \le u - p, \tag{23}$$

$$\sum_{k=1}^{N_t} |f_{ik}| b_k \le p - l, \tag{24}$$

to meet the brightness control of LEDs, and we can rewrite Eqs. (23) and (24) as

$$abs(\boldsymbol{F})\boldsymbol{b} \leq min\{\boldsymbol{u}-\boldsymbol{p}, \boldsymbol{p}-\boldsymbol{l}\}, \tag{25}$$

where $\boldsymbol{u} = u \times \mathbf{1}$, $\boldsymbol{l} = l \times \mathbf{1}$. The DC bias vector can affect the performance of precoder and equalizer, and from Ref. [13], we can know that the BER performance is best when the DC bias is the midpoint of l and u. That is, when $p = \frac{1}{2}(l+u)$, we can get a high-performance precoder and equalizer under the constraint of Eq. (25).

4.1.2. Receiver side

From **Figure 6**, we know that N_r PDs are used to convert the optical signals to electrical signals at receiver. The received signals can be expressed as

$$\boldsymbol{r} = \boldsymbol{H}\boldsymbol{y} + \boldsymbol{n} = \boldsymbol{H}(\boldsymbol{F}\boldsymbol{s} + \boldsymbol{p}) + \boldsymbol{n}, \tag{26}$$

where $\boldsymbol{n}$ is the additive white Gaussian noise (AWGN) with zero mean and independent of the data. That is, all the noise components are independently identically distributed. After removing DC components, the signal can be represented by

$$\boldsymbol{z} = \boldsymbol{r} - \boldsymbol{H}\boldsymbol{p} = \boldsymbol{H}\boldsymbol{F}\boldsymbol{s} + \boldsymbol{n}, \tag{27}$$

and a linear equalizer is used to recover the source data, which is shown as

$$\boldsymbol{d} = \boldsymbol{G}\boldsymbol{z} = \boldsymbol{G}\boldsymbol{H}\boldsymbol{F}\boldsymbol{s} + \boldsymbol{G}\boldsymbol{n}. \tag{28}$$

Finally, after the decoding part, the source signal is demodulated from $\boldsymbol{d}$. There are some detection methods to estimate the source data, and we utilize the MMSE criterion:

$$\tilde{\boldsymbol{s}} = \arg\min_{s \in S} \|\boldsymbol{d} - \boldsymbol{s}\|^2, \tag{29}$$

where S is the set of all possible source data vectors.

4.2. Transceiver design of MIMO DCO-OFDM

In this section, we focus on designing the MIMO transceiver with the knowledge of the channel state information (CSI), which means, given the channel matrix $\boldsymbol{H}$, we should determine the procoder matrix $\boldsymbol{F}$ and the equalizer matrix $\boldsymbol{G}$.

4.2.1. Iterative MMSE transceiver

In order to recover the source data from the received signal, we can formulate the problem below: design the procoder matrix $\boldsymbol{F}$ at the transmitter and the equalizer matrix $\boldsymbol{G}$ at the receiver to minimize the MSE between the transmitted data and the recovered data, that is,

$$\begin{aligned} &\min_{F,G} MSE(\boldsymbol{d}, \boldsymbol{s}, \boldsymbol{F}, \boldsymbol{G}) \\ &\text{s.t. abs}(\boldsymbol{F})\boldsymbol{b} \leq min\{\boldsymbol{u} - \boldsymbol{p}, \boldsymbol{p} - \boldsymbol{l}\}\ . \end{aligned} \tag{30}$$

Given the channel matrix $\boldsymbol{H}$, we have

$$\begin{aligned} MSE(\boldsymbol{d}, \boldsymbol{s}, \boldsymbol{F}, \boldsymbol{G}) &= \mathrm{E}\{\|\boldsymbol{d} - \boldsymbol{s}\|^2\} = \mathrm{E}\{\|\boldsymbol{GHFs} + \boldsymbol{Gn} - \boldsymbol{s}\|^2\} \\ &= E\Big\{Tr\{(\boldsymbol{GHFs} + \boldsymbol{Gn} - \boldsymbol{s})(\boldsymbol{GHFs} + \boldsymbol{Gn} - \boldsymbol{s})^H\}\Big\} \\ &= E\Big\{Tr\{[(\boldsymbol{GHF} - \boldsymbol{I})\boldsymbol{s} + \boldsymbol{Gn}][\boldsymbol{s}^H(\boldsymbol{GHF} - \boldsymbol{I})^H + \boldsymbol{n}^H\boldsymbol{G}^H]\}\Big\} \\ &= Tr\{\boldsymbol{GHFR_s F^H H^H G^H}\} + Tr\{\boldsymbol{GR_n G^H}\} + Tr\{\boldsymbol{R_s}\} - Tr\{\boldsymbol{GHFR_s}\} - Tr\{\boldsymbol{R_s F^H H^H G^H}\}, \end{aligned} \tag{31}$$

where $\boldsymbol{R_s} = \mathrm{E}\{\boldsymbol{ss}^H\}$, $\boldsymbol{R_n} = \mathrm{E}\{\boldsymbol{nn}^H\}$. In this chapter, we assume that the source data is independent of the noise. Combining Eq. (30) with Eq. (31), an iterative algorithm was developed in [13, 14], which is briefly shown in next two steps:

1. Updating $\boldsymbol{G}$ as $\boldsymbol{F}$ is given:

   Given $\boldsymbol{F}$, by setting $\frac{\partial MSE(\boldsymbol{d}, \boldsymbol{s}, \boldsymbol{F}, \boldsymbol{G})}{\partial \boldsymbol{G}} = 0$, we have the MMSE equalizer:

$$\boldsymbol{G} = \boldsymbol{R_s F^H H^H}(\boldsymbol{HFR_s F^H H^H} + \boldsymbol{R_n})^{-1}. \tag{32}$$

2. Updating $\boldsymbol{F}$ as $\boldsymbol{G}$ is given:

   Given $\boldsymbol{G}$, we can update $\boldsymbol{F}$ by minimum Eq. (31), which is linear and contains the element-wise absolute operator, and it is difficult to use the traditional optimization algorithm to get $\boldsymbol{F}$. With the help of matrix multiplication knowledge [18], the objective function of Eq. (31) can be rewritten as

$$\min_{\boldsymbol{F}} MSE(\boldsymbol{d}, \boldsymbol{s}, \boldsymbol{F}, \boldsymbol{G}) = \min_{\boldsymbol{F}} \mathrm{vec}(\boldsymbol{F}^H)^H((\boldsymbol{H}^H\boldsymbol{G}^H\boldsymbol{GH}) \otimes \boldsymbol{R_s})\mathrm{vec}(\boldsymbol{F}^H) - 2\mathrm{vec}(\boldsymbol{F}^H)^H\mathrm{vec}(\boldsymbol{R_s GH}). \tag{33}$$

So far, the original optimization problem can be transformed into a convex linearly constrained quadratic programme (LCQP) problem, which can be solved by a software package for convex programmes called CVX [13]. Through simulations, it is observed that the iterative MMSE transceiver converges after only five iterations for most times. Therefore, five iterations are involved for the iterative MMSE transceiver.

4.2.2. Advanced MMSE transceiver

The iterative MMSE transceiver presented in the previous section is a quite smart method to design the MIMO transceiver, which can effectively support the dimming control, dynamic-range constraint and other VLC-specific requirements. However, we find that the boundary condition of Eq. (25) is too harsh for practical applications, which can lead to a waste of illumination bandwidth. Thus, we propose a coefficient r to enlarge $\boldsymbol{F}$, that is,

$$\boldsymbol{F}_{adv} = \boldsymbol{F}_{ite} * r, \tag{34}$$

where $\boldsymbol{F}_{ite}$ is the precoder we get from the iterative MMSE transceiver [14] and $\boldsymbol{F}_{adv}$ is the advanced precoder. With r, the elements of $\boldsymbol{F}_{adv}$ make any signal that satisfies

$$\max_{1 \le i \le N_t} \left| \sum_{k=1}^{N_t} f_{ik} s_k \right| = \min\{p - l, u - p\}, \tag{35}$$

which improves the BER performance of MIMO transceiver in practical VLC systems.

Algorithm 2: Advanced algorithm for joint design of *F* and *G*

Step 1. Initialization:

1. Estimate channel matrix $\boldsymbol{H}$ and noise parameter $\boldsymbol{R}_n$;
2. Set illumination constraint $[l, u]$ and modulation scheme ($\boldsymbol{b}$ and $\boldsymbol{F}_s$).
3. Initialize precoder matrix $\boldsymbol{F}_0$ within the constraint of Eq. (25), for example, $\boldsymbol{F}_0 = t\boldsymbol{I}_{N_t}$, where $0 < t \le \min_{1 \le k \le N_t}\left\{\frac{u-p}{b_k}, \frac{p-l}{b_k}\right\}$.

Step 2. Iteration:

1. Update $\boldsymbol{G}_i$ with given $\boldsymbol{F}_{i-1}$ by using Eq. (32).
2. With $\boldsymbol{G}_i$, rewrite objective function Eq. (33) and figure out optimal $\boldsymbol{F}_i$ by CVX.
3. $i = i + 1$.

Step 3. Termination:

$\frac{\|\boldsymbol{F}_i - \boldsymbol{F}_{i-1}\|^2}{\|\boldsymbol{F}_{i-1}\|^2} \le 10^{-3}$, where 10^{-3} is the predefined threshold or $i = 50$, where 50 is the predefined default max iteration number.

Step 4. Application:

1. Use $\boldsymbol{F}_{ite}$ that we get from Step 3 and Eq. (35); then we can obtain r and $\boldsymbol{F}_{adv}$;
2. Update $\boldsymbol{G}_{adv}$ with $\boldsymbol{F}_{adv}$ again by using Eq. (32).
3. Apply $\boldsymbol{F}_{adv}$ and $\boldsymbol{G}_{adv}$ as MIMO transceivers in practical VLC systems.

4.3. Numerical results of MIMO DCO-OFDM

Here, we assume the channels are flat for simplicity but the results could be extended to frequency-selective channels with proper modifications. In the previous sections, we have performed theoretical analysis of the optimal transceiver design for the MIMO VLC systems, and the advanced iterative MMSE transceiver can be easily implemented into the practical design of the procoder and equalizer. In this section, we would like to show some simulation results to verify the proposed MIMO VLC system and the advanced iterative MMSE algorithm. Through simulations, it is observed that the advanced MMSE transceiver converges after only 5 iterations for most times, which is the same as the iterative MMSE transceiver. In the advanced

http://dx.doi.org/10.5772/intechopen.68887

MMSE transceiver, the number of iterations is no more than M, and in each iteration, a matrix inversion and a CVX operation are needed to be done so the total complexity of the proposed algorithm is $O(MN_t^3) + O(Q)$, and 0(Q) is the complexity of the CVX operation.

In the simulation, we consider a 5×5 MIMO VLC system and we choose the channel matric as $\boldsymbol{H} = \begin{bmatrix} 0.6386 & 0.0483 & 0.7141 & 0.4211 & 0.1484 \\ 0.4979 & 0.7962 & 0.0981 & 0.9548 & 0.0805 \\ 0.9216 & 0.9709 & 0.9140 & 0.1895 & 0.7115 \\ 0.3472 & 0.7417 & 0.7847 & 0.2734 & 0.0700 \\ 0.0221 & 0.2788 & 0.1518 & 0.0538 & 0.5241 \end{bmatrix}$; the dimming control of LEDs is selected as *l*=*1* and *u*=*9*. We assume the signals and the noise are independently identically distributed; then, we have $\boldsymbol{R}_s = \sigma_s^2 \boldsymbol{I}_5$ and $\boldsymbol{R}_n = \sigma_n^2 \boldsymbol{I}_5$. Because the precoder will adaptively adjust the transmitted signal power, we set $\sigma_s^2 = 1$. Additionally, 2-PAM/4-PAM/8-PAM/16-PAM are used, thus, $b_k = 1/\frac{3}{\sqrt{5}}/\frac{7}{\sqrt{21}}/\frac{15}{\sqrt{85}}$ (power normalization, e.g. $\sigma_s^2 = 1$). From Ref. [13], we choose $p = \frac{1}{2}(l + u) = 5$, and we can get a high-performance precoder and equalizer under the constraint of Eq. (25). For the SNR consideration, the noise power is selected from the range $[-30, 10]$ dBm.

Two different MIMO schemes are compared in our simulations: the first scheme utilizes the MMSE precoder and equalizer derived from the iterative algorithm in Section 5.2.1. In the second scheme, the advanced MMSE precoder and equalizer derived from the algorithm 2 are used. Moreover, we take into consideration the impact of dimming control and BER that is simulated as the metric to evaluate the performance. The simulation results of the BER performance of the different transceivers with various noise levels in MIMO VLC systems are shown in **Figure 7**.

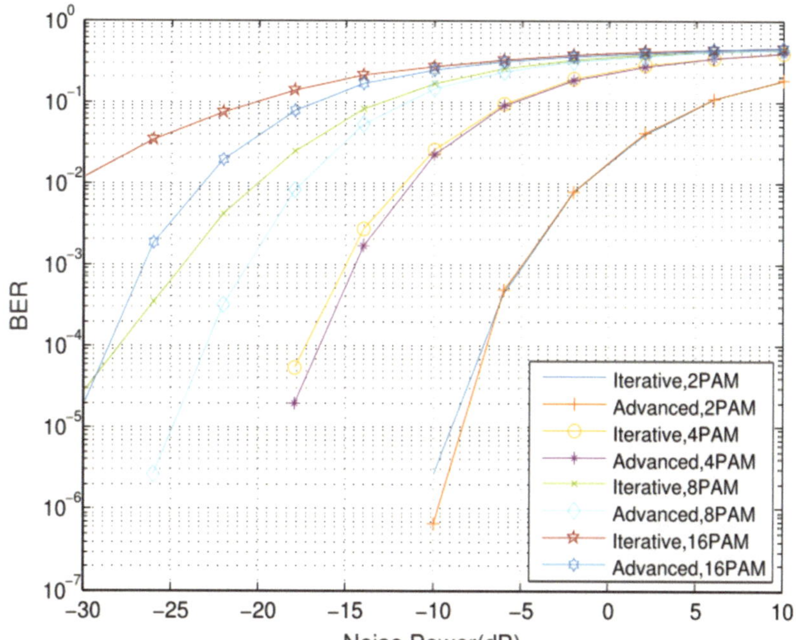

Figure 7. BER performance of the iterative MMSE transceivers and the advanced MMSE transceivers with various noise levels.

From **Figure 7**, we can see that the advanced MMSE transceiver is better than the iterative MMSE transceiver, especially when we use higher-order PAM. In fact, although the advanced MMSE transceiver is a little complicated than the iterative MMSE transceiver, it is still acceptable because the BER performance has been significantly improved and the indoor VLC system is static in most cases. The optical channel matrix does not change frequently, so the transceivers calculated from our advance MMSE algorithm can be adopted for a long time.

5. Conclusion

In this research, a novel receiver for DCO-OFDM is proposed. The receiver can accurately reconstruct the clipping noise in an iterative manner and then subtract it from the received signal, which can greatly decrease the effects of clipping noise on the system performance. Simulation results showed that the proposed receiver could achieve significant performance gain over the conventional receiver. Based on this, we investigated a MIMO VLC system with illumination control. In contract to radio frequency (RF) systems, VLC systems are limited in dynamic range, which means the transmitter should guarantee that the signal is above the turn-on value l and below the saturation value u of the LED. Under these VLC-specific requirements, we propose an advanced MMSE transceiver based on the iterative MMSE transceiver [14]. From our analysis and simulation, we conclude that the BER performance of our advanced MMSE transceiver is better than the iterative MMSE transceiver in MIMO VLC system when the DC bias is the midpoint of the dynamic range [13].

Acknowledgements

This work is supported by Southeast University 3-Category Academic Programmes Project (optical wireless communication and software-defined radio), Jiangsu NSF project (no. BK20140646), and Top-Notch Academic Programmes Project of Jiangsu Higher Education Institutions (no. PPZY2015A035).

Author details

Jian Dang*, Mengting Wu, Liang Wu and Zaichen Zhang

*Address all correspondence to: newwanda@seu.edu.cn

School of Information Science and Engineering, Southeast University, Nanjing, Peoples Republic of China

References

[1] Wang CX. Cellular architecture and key technologies for 5G wireless communication networks. IEEE Communications Magazine. 2014;**52**(2):122–130. DOI: 10.1109/MCOM.2014.6736752

[2] Komine T, Nakagawa M. Fundamental analysis for visible-light communication system using LED lights. IEEE Transactions on Consumer Electronics. 2004;**50**(1):100–107. DOI: 10.1109/TCE.2004.1277847

[3] Armstrong J. OFDM for optical communications. Journal of Lightwave Technology. 2009;**27**(3):189–204. DOI: 10.1109/JLT.2008.2010061

[4] Armstrong J, Lowery A. Power efficient optical OFDM. Electronics Letters. 2006;**42**(6):370–372. DOI: 10.1049/el:20063636

[5] Mesleh R, Elgala H, Haas H. On the performance of different OFDM based optical wireless communication systems. IEEE/OSA Journal of Optical Communications and Networking. 2011;**3**(8):620–628. DOI: 10.1364/JOCN.3.000620

[6] Armstrong J, Schmidt B. Comparison of asymmetrically clipped optical OFDM and DC-biased optical OFDM in AWGN. IEEE Communications Letters.2008;**12**(5):343–345. DOI: 10.1109/LCOMM.2008.080193

[7] Dimitrov S, Sinanovic S, Haas H. Clipping noise in OFDM-based optical wireless communication systems. IEEE Transactions on Communications. 2012;**60**(4):1072–1081. DOI: 10.1109/TCOMM.2012.022712.100493

[8] Zhang M, Zhang Z. An optimum DC-biasing for DCO-OFDM system. IEEE Communications Letters. 2014;**18**(8):1351–1354. DOI: 10.1109/LCOMM.2014.2331068

[9] Yu Z, Baxley RJ, Zhou GT. EVM and achievable data rate analysis of clipped OFDM signals in visible light communication . EURASIP Journal on Wireless Communications and Networking. 2012;**2012**(321). 1–24. DOI: 10.1186/1687-1499-2012-321

[10] Mesleh R, Mehmood R, Elgala H, Haas H. Indoor MIMO optical wireless communication using spatial modulation. In: IEEE International Conference on Communications; 23-27 May 2010; Cape Town, South Africa. IEEE; 2010. 1–5. DOI: 10.1109/ICC.2010.5502062

[11] Azhar AH, Tran TA, O'Brien D. Demonstration of high-speed data transmission using MIMO-OFDM visible light communications. In: IEEE Globecom Workshop OWC; 6-10 December 2010; Miami, Florida. USA: IEEE; 2011. 1052–1056. DOI: 10.1109/GLOCOMW.2010.5700095

[12] Azhar AH, Tran T-A, O'Brien DC. A gigabit/s indoor wireless transmission using MIMO-OFDM visible-light communications. IEEE Photonics Technology Letters. 2013;**25**(2):171–174. DOI: 10.1109/LPT.2012.2231857

[13] Ying K, Qian H, Baxley RJ, Yao S. Joint optimization of precoder and equalizer in MIMO VLC systems. IEEE Journal on Selected Areas in Communications. 2015;**33**(9):1949–1958. DOI: 10.1109/JSAC.2015.2432515

[14] Ying K, Qian H, Baxley RJ, Tong Zhou G. MIMO transceiver design in dynamic-range-limited VLC systems. IEEE Photonics Technology Letters. 2016;**28**(22):2593–2596. DOI: 10.1109/LPT.2016.2606341

[15] Burton A, Minh HL, Ghassemlooy A, Bentley E, Botella C. Experimental demonstration of 50-Mb/s visible light communications using 4×4 MIMO. IEEE Photonics Technology Letters. 2014;**26**(9):945–948. DOI: 10.1109/LPT.2014.2310638

[16] Thao PC, Khoa DL, Tu NT, Phuc LH, Phuong NH. Optical MIMO DCO-OFDM wireless communication systems using STBC in diffuse fading channels. In: National Foundation for Science and Technology Development Conference on Information and Computer Science; 14-16 September 2016. IEEE; 2016. 141–146. DOI: 10.1109/NICS.2016.7725639

[17] Stepniak G, Siuzdak J, Zwierko P. Compensation of a VLC phosphorescent white LED nonlinearity by means of Volterra DFE. IEEE Photonics Society. 2010;**25**(16):1597–1600. DOI: 10.1109/LPT.2013.2272511

[18] Horn RA, Johnson CR, editors. Topics in Matrix Analysis. New York, USA: Cambridge University Press; 1991. p. 616. DOI: 10.1137/1035037

Chapter 4

Index Modulation-Aided OFDM for Visible Light Communications

Qi Wang, Tianqi Mao and Zhaocheng Wang

Additional information is available at the end of the chapter

http://dx.doi.org/10.5772/intechopen.68885

Abstract

Index modulation-aided orthogonal frequency-division multiplexing(IM-OFDM) is a promising modulation technique to achieve high spectral and energy efficiency. In this chapter, the conventional optical OFDM schemes are firstly reviewed, followed by the principles of IM-OFDM. The application of IM-OFDM in visible light communication (VLC) systems is introduced, and its performance is compared with conventional optical OFDM, which verifies its superiority. Finally, the challenges and opportunities of IM-OFDM are discussed for the VLC applications.

Keywords: index modulation (IM), orthogonal frequency-division multiplexing (OFDM), visible light communications (VLCs), spectral efficiency, energy efficiency

1. Introduction

Orthogonal frequency-division multiplexing (OFDM) has become a ubiquitous digital communication technique, which is widely employed in visible light communication (VLC) [1]. Since intensity modulation with direct detection is utilized in VLC for low-cost implementation, the signals modulated on the light-emitting diodes (LEDs) should be real-valued and nonnegative [2]. Therefore, various optical OFDM schemes have been proposed to satisfy these constraints, namely DC-biased optical OFDM (DCO-OFDM) [3], asymmetrically clipped optical OFDM (ACO-OFDM) [4], pulse-amplitude-modulated discrete multi-tone (PAM-DMT) [5], unipolar OFDM (U-OFDM) [6], and Flip OFDM [7]. In all these optical OFDM schemes, Hermitian symmetry is utilized on the OFDM subcarriers before inverse fast Fourier transform (IFFT), so that the time-domain signals are real-valued. To ensure the nonnegativity, DC bias can be imposed on the resultant signals, leading to low-energy efficiency. Alternatively, special arrangements on the signals in the time or frequency domain are used in Refs. [3–7] without the need of DC bias, which generate non-negative signals at the cost of spectral efficiency loss. In order to overcome the spectral efficiency loss while maintaining high-energy efficiency, several hybrid schemes are proposed [8–11]. For example, inasymmetrically clipped DC-biased optical OFDM (ADO-OFDM) [8] and hybrid ACO-OFDM (HACO-OFDM) [9], ACO-OFDM with odd subcarriers is combined with DCO-OFDM and PAM-DMT modulating even subcarriers for simultaneous transmission, respectively. In layered ACO-OFDM (LACO-OFDM) [10] and enhanced U-OFDM (eU-OFDM) [11], multiple streams of ACO-OFDM or U-OFDM are superposed for higher spectral efficiency.

Recently, the index modulation (IM) technique has been introduced to OFDM to enhance its performance [12, 13], where the information is transmitted not only with the classic amplitude and phase modulation schemes but also implicitly by the indices of the activated subcarriers. Compared with classical OFDM, IM-aided OFDM (IM-OFDM) offers a promising trade-off between the bit error rate (BER) performance and the spectral efficiency by changing the number of activated subcarriers, the usage of constellation alphabets, and so on. In fact, the primary concept of IM-aided OFDM was proposed two decades ago, which was termed as parallel combinatory OFDM [14]. After the development of spatial modulation (SM) (or IM) in recent years, it attracts extensive attentions. In Refs. [15], subcarrier index modulation (SIM-OFDM) is proposed, where the status of each subcarrier (ON or OFF) carries one-bit information while the activated subcarriers are modulated by conventional constellations such as quadrature amplitude modulation (QAM) and phase shift keying (PSK). Additional bits can be transmitted by the indices of activated subcarriers, and the energy efficiency is improved with inactivated subcarriers. However, since the bit rate is unstable for different information bits, it may cause error propagation at the receiver. To address this issue, enhanced SIM-OFDM (ESIM-OFDM) is proposed in Refs. [16], where every two subcarriers are paired and only one subcarrier is activated in each pair. However, the spectral efficiency is reduced since the bits transmitted by indices of subcarriers are halved. In Refs. [17], OFDM with index modulation (OFDM-IM) is proposed where subcarriers are partitioned into several subblocks and the information bits are transmitted by both the indices of activated subcarriers and signal constellations in each subblock. The benefit of OFDM-IM is that less power is required since only a fraction of the subcarriers is employed for modulation, while the indices of activated subcarriers can be utilized to transmit extra bits. However, the inactivated subcarriers waste enormous precious frequency resources and reduce the spectral efficiency significantly. Moreover, it is shown that although OFDM-IM outperforms conventional OFDM with low spectral efficiency below 2 bit/s/Hz, it may perform even worse than OFDM when high-order constellations are utilized to achieve high spectral efficiency [18]. Furthermore, a subcarrier-level interleaving technique is introduced to OFDM-IM in Refs. [19], which enlarges the Euclidean distances between different transmitted symbols. In Refs. [20], OFDM-IM is combined with space-time block codes with coordinate interleaving, enhancing its BER performance due to the additional diversity gain. Besides, generalized schemes of OFDM-IM have been proposed in Refs. [21, 22], where the number of activated subcarriers of each OFDM subblock is variable, and index modulation is performed on both the in-phase and quadrature components of the modulated symbols, respectively. The generalized schemes are capable of enhancing the spectral efficiency of conventional OFDM-IM significantly. Additionally, [23] facilitates a spectrally efficient IM-OFDM scheme by employing various constellations and numbers of activated subcarriers in different subblocks. Moreover, OFDM-IM is also combined with multiple-input-multiple-output (MIMO) systems [24, 25], yielding considerable performance improvement over conventional MIMO-OFDM. Furthermore, OFDM-IM has been applied to underwater acoustic communications as well as vehicle-to-vehicle and vehicle-to-infrastructure applications [26–28], leading to performance gains. Recently, the dual-mode index modulation-aided OFDM (DM-OFDM) is proposed in Refs. [29], where all the subcarriers are utilized to carry information unlike OFDM-IM. For each subblock, the subcarriers are divided into two groups modulated by two different constellation modes. The indices of either group can be used to transmit extra information bits. Therefore, DM-OFDM achieves higher spectral efficiency than both OFDM-IM and conventional OFDM. The generalized scheme of DM-OFDM is invoked in Refs. [30] to further enhance the spectral efficiency, where the number of subcarriers modulated by the same constellation mode is alterable. In addition, the performance trade-off of IM-OFDM schemes is investigated with theoretical analysis in Refs. [31, 32], which help to enhance the overall performance of IM-OFDM.

Due to the distinct advantages of IM-OFDM, it is also applied to VLC systems. Considering the intensity modulation property of VLC, the existent IM-OFDM schemes cannot be directly introduced to VLC, and several optical IM-OFDM (O-IM-OFDM) schemes [33, 34] have been proposed. Hence, O-IM-OFDM is investigated in this chapter by introducing the transceiver design, performing the theoretical analysis, evaluating the numerical results, and pointing out its challenges and potentials.

The rest of this chapter is organized as follows. Section 2 reviews the optical OFDM schemes for VLC, while the principles of IM-OFDM for VLC are detailed in Section 3. In Section 4, several challenges and opportunities are discussed for the deployment of IM-OFDM in VLC, and conclusions are drawn in Section 5.

2. Optical OFDM for visible light communications

Due to the intensity modulation property, the transmitted signals are constrained to be real-valued and nonnegative in VLC. Typically, Hermitian symmetry is imposed on the frequency-domain subcarriers to generate real outputs after the IFFT, and the symbols modulated onto the subcarriers satisfy

$$X_k = X^*_{N-k}, \; k = 1, 2, \ldots, N/2 - 1, \tag{1}$$

where N is the number of subcarriers in OFDM. Moreover, X_0 and $X_{N/2}$ are set to zero for the same purpose. After the IFFT, the resultant time-domain signal can be formulated as

$$x_n = \frac{1}{\sqrt{N}} \sum_{k=0}^{N-1} X_k \exp\left(j\frac{2\pi}{N} nk\right), \; n = 0, 1, \ldots, N-1, \tag{2}$$

which is bipolar, and various techniques have been proposed to obtain nonnegative waveforms for transmission. When $N \geq 64$, the distribution of x_n is approximately Gaussian with zero mean. In the following, DCO-OFDM, ACO-OFDM, PAM-DMT, U-OFDM, and hybrid optical OFDM schemes are introduced, and the diagram of these schemes is illustrated in **Figure 1**.

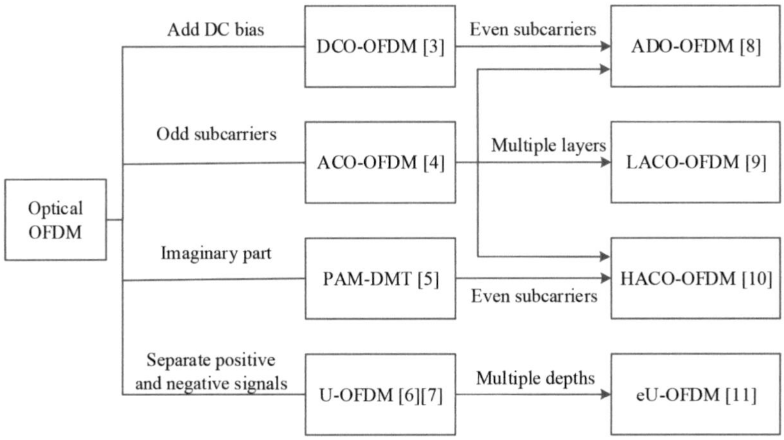

Figure 1. Diagram of optical OFDM schemes for VLC.

2.1. DCO-OFDM

A natural way to obtain the unipolar signal is adding a DC bias to the bipolar signal while clipping the remaining negative signal at zero. We denote the DC bias as x_{DC}, which is usually proportional to the square root of the electric power of x_n in Eq. (2). Since the expectation of x_n is zero, the optical power of the transmitted signal is proportional to the DC bias when clipping distortion is neglected. A large value of x_n leads to small clipping distortion, which is beneficial to the BER performance at the receiver. However, DC bias does not carry useful information, which is not energy efficient. Therefore, a trade-off between clipping distortion and energy efficiency should be made [35].

2.2. ACO-OFDM

In order to avoid DC bias to achieve higher energy efficiency, ACO-OFDM is proposed, which leaves the even subcarriers unmodulated to obtain an anti-symmetric waveform. For each positive time-domain signal, there is a negative signal with the same absolute value at certain position. Specifically, the time-domain signals of ACO-OFDM after the IFFT satisfy [4]

$$x_{\mathrm{ACO},n} = -x_{\mathrm{ACO},n+N/2}, \quad n = 0, 1, \ldots, N/2 - 1, \tag{3}$$

whose negative part can be directly clipped at zero (*asymmetrically clipped*) without information loss. Therefore, the transmitted signal of ACO-OFDM is written as

$$x^c_{\mathrm{ACO},n} = x_{\mathrm{ACO},n} + i_{\mathrm{ACO},n} = \begin{cases} x_{\mathrm{ACO},n}, & x_{\mathrm{ACO},n} \geq 0; \\ 0, & x_{\mathrm{ACO},n} < 0 \end{cases} \tag{4}$$

For $n = 0,1,\ldots, N-1$, where $i_{\mathrm{ACO},n}$ is the negative clipping distortion of ACO-OFDM. Interestingly, the FFT of $i_{\mathrm{ACO},n}$ denoted as $I_{\mathrm{ACO},k}$ only falls on the even subcarriers, which does not interfere with the useful information. Therefore, a simple FFT can be used at the receiver for detection.

2.3. PAM-DMT

PAM-DMT is also an asymmetrically clipped method to generate nonnegative OFDM signals. Unlike ACO-OFDM, the imaginary part of all subcarriers is utilized for data transmission, while the real part is left unused to produce an anti-symmetric waveform. For the imaginary part of each subcarrier, PAM constellations are employed. The time-domain signals of PAM-DMT after the IFFT satisfy

$$x_{\mathrm{PAM},n} = -x_{\mathrm{PAM},N-n}, \quad n = 1, \ldots, N/2 - 1, \tag{5}$$

and the transmitted signal of PAM-DMT denoted as $x^{c}_{\mathrm{PAM},n}$ can be obtained similar to Eq. (4), while we denote the negative clipping distortion of PAM-DMT as $i_{\mathrm{PAM},n}$. The FFT of $i_{\mathrm{PAM},n}$ represented by $i_{\mathrm{PAM},k}$ only falls on the real part of each subcarrier, which does not interfere with the useful information similar to ACO-OFDM. Therefore, a simple FFT can be employed for detection at the receiver.

2.4. U-OFDM

In U-OFDM, the signal in Eq. (2) is utilized to generate nonnegative signal. Unlike DCO-OFDM where the DC bias is used to make the signal unipolar, the OFDM frame is separated into two frames with the same length. In the first frame, all the positive signals remain the same, while the negative signals are clipped at zero. In the second frame, all the negative signals are replaced by their absolute values, while the positive signals are set to zero. At the receiver, the second frame is subtracted from the first frame to recover the original signal. Afterward, the FFT can be used to the resultant signal for detection similar to DCO-OFDM.

2.5. Hybrid optical OFDM

It is noted that, although ACO-OFDM, PAM-DMT, and U-OFDM do not require DC bias, only half of the resources (in either frequency domain or time domain) are employed compared with DCO-OFDM. Therefore, their performances are better than DCO-OFDM only when low-order constellations are used [8]. When high spectral efficiency is required, DCO-OFDM is more preferred since all the resources are used. Recently, some hybrid optical OFDM schemes have been proposed to achieve better trade-off between spectral efficiency and energy efficiency.

In ADO-OFDM, the ACO-OFDM signal is superposed by the DCO-OFDM signal, where only the even subcarriers are modulated by DCO-OFDM to avoid the interference [8]. In this way, all the subcarriers are utilized for modulation, leading to improved spectral efficiency. At the receiver, the symbols on the odd subcarriers for ACO-OFDM are firstly demodulated after the FFT. In order to detect the symbols on the even subcarriers for DCO-OFDM, the clipping distortion of ACO-OFDM should be eliminated firstly, which is estimated by the recovered symbols of ACO-OFDM. However, DC bias is still required in ADO-OFDM (although the power is reduced), which is inefficient in terms of power.

In HACO-OFDM, ACO-OFDM is combined with PAM-DMT whose even subcarriers are modulated by PAM [10]. The odd subcarriers in PAM-DMT are also left unused as in ADO-OFDM.

Therefore, it is very similar to ADO-OFDM at both the transmitter and the receiver. Since PAM-DMT does not require DC bias, HACO-OFDM is more energy-efficient compared with ADO-OFDM. However, the real part of even subcarriers is unmodulated, thus its spectral efficiency is still limited.

To fully utilize the frequency resources, LACO-OFDM is proposed in Ref. [9], where multiple layers of ACO-OFDM are combined for simultaneous transmission. Different layers employ different subcarriers to generate the nonnegative ACO-OFDM signals, while successive interference cancellation is used at the receiver to recover all the symbols in each layer. Compared with conventional ACO-OFDM, the spectral efficiency of LACO-OFDM is approximately doubled with the same constellation order. A similar method is applied to U-OFDM called eU-OFDM, which combines different depths of U-OFDM to further improve the spectral efficiency [11]. In different depths, various repetitions are required so that they can be recovered at the receiver.

3. Index modulation-aided OFDM for visible light communications

In this section, two representative IM-OFDM schemes, namely OFDM-IM and DM-OFDM, will be investigated in terms of their principles and applications in VLC.

3.1. Principles of IM-OFDM

The basic idea of OFDM-IM is to divide the subcarriers into several groups and the indices of activated subcarriers in each group can be used to convey extra information. We denote the number of subcarriers in OFDM as N, while m bits are transmitted within one OFDM symbol. The m bits are split into g groups each consisting of p bits, where we have $p = m / g$. Besides, the subcarriers are divided into g subblocks as well, and each subblock has $n = N / g$ subcarriers. In each subblock of OFDM-IM, k out of n subcarriers are activated for modulation, while the others are left empty. Therefore, the indices of the activated subcarriers can carry p_1 bits index information, which is given by

$$p_1 = \left\lfloor \log_2 \binom{n}{k} \right\rfloor, \tag{6}$$

where $\lfloor \cdot \rfloor$ denotes the integer floor operator. When M–ary constellation $\mathcal{M}$ is used for the k-activated subcarriers, p_2 bits can be transmitted by each OFDM subblock, which is given by

$$p_2 = k\log_2(M). \tag{7}$$

Therefore, the number of total transmitted bits in an OFDM-IM frame is expressed as

$$m = g(p_1 + p_2) = g\left(\left\lfloor \log_2 \binom{n}{k} \right\rfloor + k\log_2(M) \right), \tag{8}$$

and the spectral efficiency is given by

$$\gamma_{IM} = \frac{m}{N} = \frac{\left\lfloor \log_2 \binom{n}{k} \right\rfloor + k\log_2(M)}{n} \text{ bit/s/Hz.} \tag{9}$$

Take $n = 4$, $k = 2$, $M = 2$, for example. The spectral efficiency of OFDM-IM is 1 bit/s/Hz, which is identical to the spectral efficiency of conventional OFDM scheme with BPSK modulation. Since only half of the subcarriers are activated in OFDM-IM, the energy efficiency is improved significantly.

It is shown in Eq. (6) that the number of index bits in OFDM-IM is constant when n and k are fixed. When high-order constellations are used in activated symbols, the information bits provided by the index patterns are negligible for the whole spectral efficiency. The spectral efficiency loss of the inactivated subcarriers cannot be compensated by the index bits. Moreover, frequency resource is very precious, which is unfavorable to be wasted. Therefore, DM-OFDM is proposed in Ref. [29] to fully exploit the subcarriers and index information.

In DM-OFDM, two constellation sets $\mathcal{M}_A$ and $\mathcal{M}_B$ are utilized for each subblock, whose sizes are M_A and M_B, respectively. In order to ensure the detection at the receiver, the two constellations sets should have no common constellations points, that is, $\mathcal{M}_A \cap \mathcal{M}_B = \emptyset$. In each subblock, the subcarriers are divided into two groups A and B, whose index sets are $\mathbf{I}_A$ and $\mathbf{I}_B$. The symbols in $\mathcal{M}_A$ and $\mathcal{M}_B$ are used for modulation in subcarrier groups A and B, respectively. Since $\mathbf{I}_B$ can be easily determined by $\mathbf{I}_A$, we denote $\mathbf{I}_A$ as the *index pattern* of the OFDM subblock, which can be utilized to convey index bits. When k subcarriers are modulated by $\mathcal{M}_A$ in each subblock, the number of total transmitted bits of a DM-OFDM symbol is expressed as

$$m = g\left(\left\lfloor \log_2 \binom{n}{k} \right\rfloor + k\log_2(M_A) + (n-k)\log_2(M_B)\right), \tag{10}$$

and the spectral efficiency of DM-OFDM is calculated as

$$\gamma_{DM} = \frac{m}{N} = \frac{\left\lfloor \log_2 \binom{n}{k} \right\rfloor + k\log_2(M_A) + (n-k)\log_2(M_B)}{n} \text{ bit/s/Hz.} \tag{11}$$

3.2. OFDM-IM for visible light communications

OFDM-IM can be applied to VLC with some modifications. The block diagram of OFDM-IM-based VLC system [33] is illustrated in **Figure 2**. Since intensity modulation is utilized in VLC, Hermitian symmetry is required to generate real-valued signals. Therefore, only half of the subcarriers are utilized for grouping and bit mapping, while the other half of the subcarriers can be obtained by simple Hermitian symmetry. After N-point IFFT, the bipolar signals are passed through the unipolar conversion block to generate unipolar signals for LED emission. When DC-biased optical OFDM-IM (DCO-OFDM-IM) is considered, a DC bias is utilized, while the remaining negative signals are directly clipped at zero. When unipolar OFDM-IM (U-OFDM-IM) is used, the OFDM frame is separated into two frames with the same length

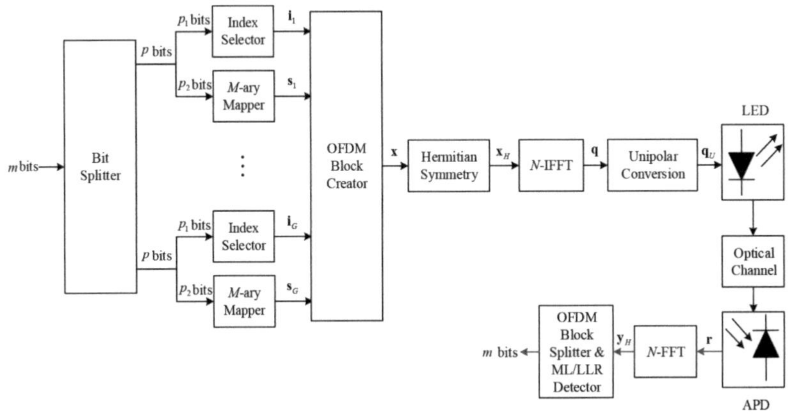

Figure 2. Block diagram of OFDM-IM-based VLC system.

similar to U-OFDM. All the positive signals are transmitted in the first frame, while the negative signals are inverted for transmission in the second frame.

Therefore, the spectral efficiencies of DCO-OFDM-IM and U-OFDM-IM are given by

$$\gamma_{\text{DCO-IM}} \approx \frac{\left\lfloor \log_2 \binom{n}{k} \right\rfloor + k\log_2(M)}{2n} \text{ bit/s/Hz}, \tag{12}$$

$$\gamma_{\text{U-IM}} = \frac{\left\lfloor \log_2 \binom{n}{k} \right\rfloor + k\log_2(M)}{4n} \text{ bit/s/Hz}. \tag{13}$$

At the receiver, if U-OFDM-IM is employed, the second frame is subtracted from the first frame. Afterward, an N-point FFT is performed on the time-domain signals. Unlike conventional optical OFDM where the detection can be performed symbol by symbol, the optical OFDM-IM requires subblock-by-subblock detection since indices of activated subcarriers are unknown at the receiver. Without loss of generality, we consider the symbols in one subblock, where the received symbols after the FFT can be written as

$$R_i = H_i X_i + W_i, i = 1, 2, \ldots, n, \tag{14}$$

where H_i, X_i, and W_i denote the channel response, transmitted symbol, and additive white Gaussian noise (AWGN) with the variance of N_0 in the frequency domain, respectively. When all the symbols are considered in the subblock, Eq. (14) can be rewritten in the vector form as

$$\mathbf{R} = \mathbf{H}\mathbf{X} + \mathbf{W}, \tag{15}$$

where the diagonal matrix $\mathbf{H} = \text{diag}(H_1, H_2, \ldots, H_g)$. The optimal receiver employs maximum likelihood (ML) detection, which considers all the possible subblock realizations with different activated subcarrier indices and the constellation points by minimizing the metric as follows:

$$\{\hat{\mathbf{I}}, \hat{\mathbf{X}}\} = \underset{\mathbf{I},\mathbf{X}}{\mathrm{argmin}} \|\mathbf{R} - \mathbf{H}\mathbf{X}\|^2, \tag{16}$$

where **I** is the index pattern of the activated subcarriers. The ML detector should consider $2^{\left\lfloor \log_2 \binom{n}{k} \right\rfloor + k\log_2(M)}$ possible realizations, whose complexity is very high for large values of n, k, and M. Therefore, low-complexity detection is required for practical implementation.

In order to reduce the complexity of the receiver, log-likelihood ratio (LLR) detector is proposed by calculating the logarithm of the ratio between *a posteriori* probabilities of the frequency-domain symbols being either nonzero or zero. A larger LLR means it is more likely that the corresponding subcarrier is activated. The LLR for the i-th subcarrier is given by

$$\eta(i) = \mathbf{log}\left(\frac{\sum_{t=1}^{M} \mathbf{Pr}(X_i = S_t | R_i)}{\mathbf{Pr}(X_i = \mathbf{0} | R_i)} \right), \quad \mathbf{1} \leq i \leq n, \tag{17}$$

where $S_t \in \mathcal{M}$. Since $\sum_{t=1}^{M} \Pr(X_i = S_t) = k/n$ and $\mathbf{Pr}(X_t = 0) = (n-k)/n$, the LLR can be rewritten as

$$\eta(i) = \mathbf{ln}(k) - \mathbf{ln}(n-k) + \frac{|R_i|^2}{N_0} + \mathbf{ln}\left(\sum_{t=1}^{M} \mathbf{exp}\left(-\frac{\mathbf{1}}{N_0} |R_i - H_i S_t|^2 \right) \right). \tag{18}$$

When the activated subcarriers are detected, the symbols on the subcarriers can be demodulated independently as the conventional optical OFDM. Therefore, the complexity of the LLR detector for optical OFDM-IM is similar to that of conventional optical OFDM.

3.3. DM-OFDM for visible light communications

The block diagram of DM-OFDM-based VLC system [34] is shown in **Figure 3**. Unlike other IM-OFDM schemes where several subcarriers are empty in each subblock, DM-OFDM utilizes two distinguishable constellation sets to modulate all the subcarriers, thus achieving higher spectral efficiency. The spectral efficiencies of DCO-DM-OFDM and U-DM-OFDM are given by

$$\gamma_{\mathrm{DCO-DM}} \approx \frac{\left\lfloor \log_2 \binom{n}{k} \right\rfloor + k\log_2(M_A) + (n-k)\log_2(M_B)}{2n} \ \text{bit/s/Hz}. \tag{19}$$

$$\gamma_{\mathrm{U-DM}} = \frac{\left\lfloor \log_2 \binom{n}{k} \right\rfloor + k\log_2(M_A) + (n-k)\log_2(M_B)}{4n} \ \text{bit/s/Hz}. \tag{20}$$

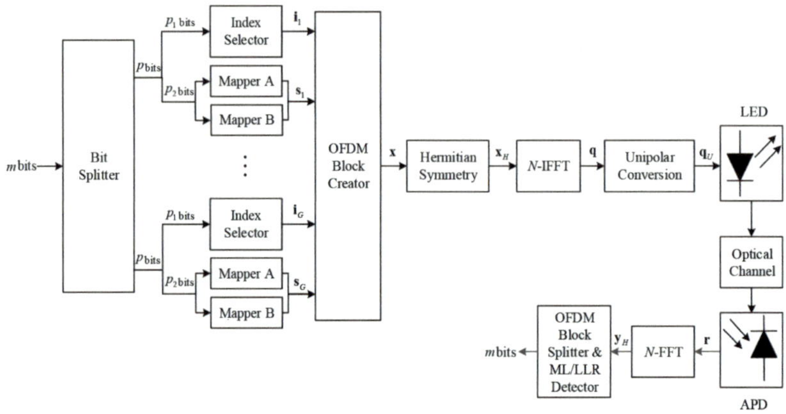

Figure 3. Block diagram of DM-OFDM for VLC systems.

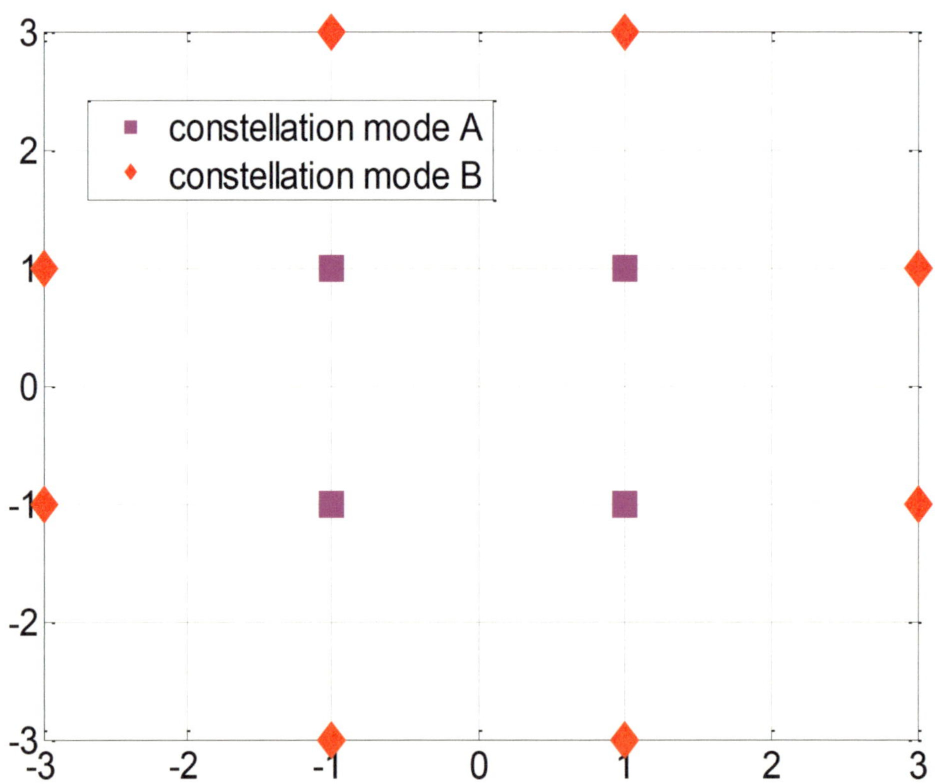

Figure 4. An example of DM-OFDM constellation design for $\mathcal{M}_A$ and $\mathcal{M}_B$ with $M_A = 8$ and $M_B = 4$.

Obviously, it is crucial to find the good combinations of two constellation sets in DM-OFDM. To ensure good BER performance, the minimum Euclidean distance between the two constellations should be equal to that of the points within each constellation. Therefore, one can firstly design a constellation with $M_A + M_B$ points and then separate the points into two constellations.

Specifically, when $M_A = M_B = M$, we can employ the points in the $2M$–ary QAM constellation. In **Figure 4**, an example of DM-OFDM constellation design is provided for $\mathcal{M}_A$ and $\mathcal{M}_B$ with $M_A = 8$ and $M_B = 4$.

The detection of DM-OFDM is similar to that of OFDM-IM in a subblock-by-subblock manner. When ML detection is used, we have

$$\{\hat{\mathbf{I}}_A, \hat{\mathbf{X}}\} = \underset{\mathbf{I}_A, \mathbf{X}}{\arg\min} \|\mathbf{R} - \mathbf{H}\mathbf{X}\|^2, \tag{21}$$

which still has high complexity.

The LLR detector for DM-OFDM is given by

$$\eta(i)=\ln\left(\frac{M_B k}{M_A(n-k)}\right)+\ln\left(\sum_{t=1}^{M_A}\exp\left(-\frac{1}{N_0}|R_i-H_i S_t^A|^2\right)\right)-\ln\left(\sum_{t=1}^{M_B}\exp\left(-\frac{1}{N_0}|R_i-H_i S_t^B|^2\right)\right), \tag{22}$$

where $S_t^A \in \mathcal{M}_A$ and $S_t^B \in \mathcal{M}_B$. Different from OFDM-IM, Eq. (22) calculates the logarithm of the ratio between the *a posteriori* probabilities of the frequency-domain symbols being modulated by mapper A and mapper B. When the index pattern is detected, the symbol on each subcarrier can be demodulated by the corresponding demapper.

3.4. Performance evaluation

The BER performances of OFDM-IM and DM-OFDM schemes are compared with their conventional optical OFDM counterparts in VLC systems. Unlike antennas in the radio frequency communication systems, LEDs are used in VLC systems, which have nonlinear transfer characteristics, leading to distortions on the transmitted signal beyond the linear range. The nonlinearity can be simply modeled as

$$T(x) = \begin{cases} V_{\min}, x < V_{\min}; \\ x, V_{\min} \le x \le V_{\max}; \\ V_{\max}, x > V_{\max}, \end{cases} \tag{23}$$

where $V_{\max}$ and $V_{\min}$ denote the maximum and minimum allowed amplitudes, respectively.

In the simulations, the size of IFFT is set to 256, and the number of subcarriers in each subblock is 4. In OFDM-IM, 2 subcarriers are activated in each subblock. While in DM-OFDM, two subcarriers are modulated by mapper A, and the other two subcarriers are modulated by mapper B. The linear range of LEDs is [0, 1], and the DC bias is set to 0.5 for DCO-OFDM, DCO-OFDM-IM, and DCO-DM-OFDM. In U-OFDM, U-OFDM-IM, and U-DM-OFDM, no DC bias is required.

Figure 5 illustrates the BER performances of OFDM-IM and the conventional optical OFDM schemes, where quadrature phase shift keying (QPSK) and binary phase shift keying (BPSK) are utilized in OFDM-IM and its conventional counterparts, respectively. In the simulations, the x-axis E_b/N_0 stands for the signal-to-noise ratio per bit. Besides, the input energies into

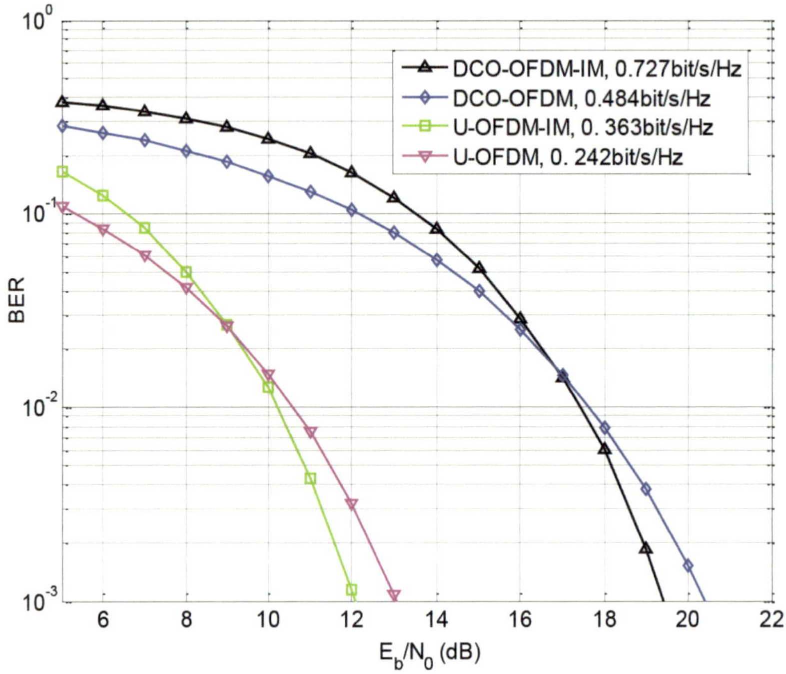

Figure 5. Performance comparison between OFDM-IM and the conventional optical OFDM schemes.

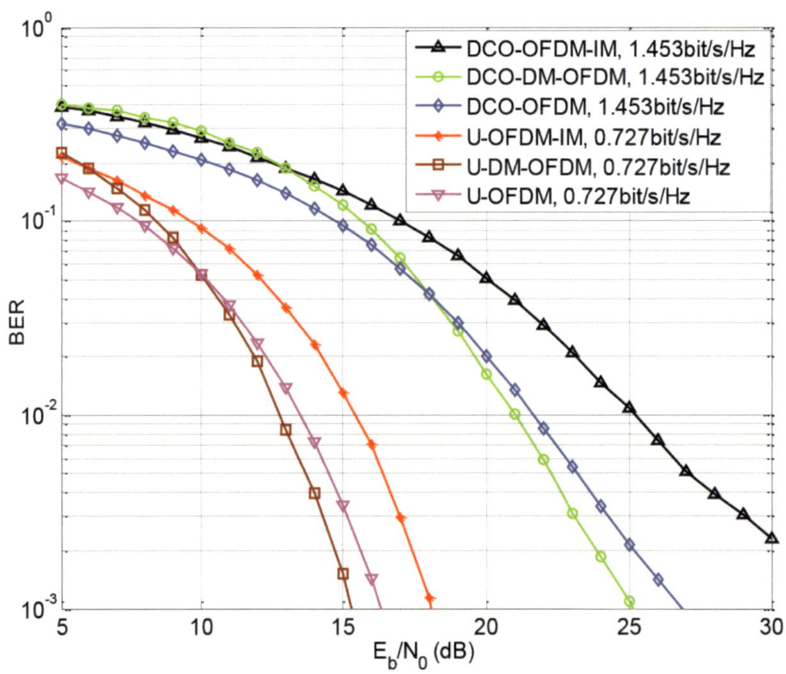

Figure 6. Performance comparison between OFDM-IM, DM-OFDM, and the conventional optical OFDM schemes at higher spectral efficiency.

LEDs are set as 13 and 13.5 dBm (not including the DC energy) for DC-based and U-based schemes, respectively. For both cases, OFDM-IM achieves about 0.243 and 0.121 bit/s/Hz spectral efficiency gains over the conventional DCO-OFDM and U-OFDM. However, it can be seen that OFDM-IM still outperforms its conventional counterparts at the BER level of 10^{-3}, since enhanced spectral efficiency leads to reduced average bit energy E_b, yielding smaller required E_b/N_0.

When higher spectral efficiency is considered, OFDM-IM might perform even worse than conventional optical OFDM. **Figure 6** shows the BER performances of OFDM-IM, DM-OFDM and the conventional optical OFDM with a spectral efficiency of 1.453 bit/s/Hz for DC-based schemes and 0.727 bit/s/Hz for U-based schemes, where the input energies into LEDs are set as 13 and 13.5 dBm (not including the DC energy) for DC-based and U-based schemes, respectively. In conventional optical OFDM schemes, 8 QAM is employed on each subcarrier, while 32 QAM is used in OFDM-IM. In DM-OFDM, the constellations utilized are shown in **Figure 4**. From **Figure 6**, it is shown that OFDM-IM suffers from performance loss compared with conventional optical OFDM with high spectral efficiency, while DM-OFDM achieves more than 1 and 2 dB performance gains over U-OFDM and DCO-OFDM at the BER of 10^{-3}, for the reason that all the subcarriers are used for modulation, and the indices of subcarriers can provide additional dimension for transmission.

4. Challenges, opportunities, and future research trends

4.1. Dimming compatibility

In VLC systems, illumination is an important function of LEDs. Therefore, the modulation scheme should be compatible with dimming control. When the required illumination level is changed, the data rate should not fluctuate too much. In addition, the modulation scheme should support extreme illumination requirements such as very low or very high intensities. Several schemes have been proposed for conventional optical OFDM to support dimming control in VLC [36, 37], which can be extended to the IM-OFDM schemes. Furthermore, given the property of IM-OFDM, we can use different number of activated subcarriers for various illumination requirements.

4.2. MIMO transmission

In indoor environments, multiple LEDs are installed to provide sufficient illumination. Therefore, it is worthwhile to study the modulation schemes in the MIMO VLC systems [38]. For MIMO transmission, low-complexity transceiver should be designed under the intensity modulation constraint. Moreover, the index modulation can be utilized in the time domain and space domain, and multi-dimensional index modulation in time, frequency, and space domains might be useful to achieve better performance [39, 40].

4.3. Performance improvement

Current researches on IM-OFDM for VLC only consider simple combinations of optical OFDM and IM. Hybrid optical OFDM schemes introduced in Section 2.5 can be considered to further improve the spectral efficiency of IM-OFDM for VLC systems. At the transmitter, IM-OFDM for VLC suffers from high peak-to-average power ratio (PAPR), causing nonlinear distortions of LEDs. Thus, efficient PAPR reduction techniques have to be employed to combat with the nonlinearity of LEDs. Besides, the low-complexity and high-performance receiver design also requires attention for the deployment of IM-OFDM in low-cost VLC systems.

5. Conclusions

In this chapter, to shed light on the development of IM-OFDM schemes in VLC, we introduce the principles of optical OFDM and IM-OFDM, which are exemplified by several representative schemes. It is indicated that various IM-OFDM schemes are capable of enhancing the energy efficiency or the spectral efficiency compared with conventional OFDM, leading to improved BER performance. Theoretical analysis and numerical results demonstrate that these IM-OFDM schemes can enhance the overall performance compared with its conventional counterparts. Therefore, IM-OFDM would be a promising modulation technique for future VLC systems. Moreover, the challenges and opportunities are discussed for the deployment of IM-OFDM in VLC systems, which are beneficial to researchers who are interested in this field.

Author details

Qi Wang[1], Tianqi Mao[2] and Zhaocheng Wang[2*]

*Address all correspondence to: zcwang@tsinghua.edu.cn

1 School of Electronics and Computer Science, University of Southampton, United Kingdom

2 Department of Electronic Engineering, Tsinghua National Laboratory for Information Science and Technology (TNList), Tsinghua University, Beijing, China

References

[1] Tsonev D, Chun H, Rajbhandari S, McKendry JJD, Gu E, Haji M, et al. A 3-Gb/s single-LED OFDM-based wireless VLC link using a gallium nitride μLED. IEEE Photonics Technology Letters. 2014;**26**(7):637–640

[2] Armstrong J. OFDM for optical communications. Journal of Lightwave Technology. 2009;**27**(3):189–204

[3] Carruthers J and Kahn J. Multiple–subcarrier modulation for nondirected wireless infrared communication. IEEE Journal Selected Areas in Communications. 1996;**14**(3):538–546

[4] Armstrong J and Lowery A. Power efficient optical OFDM. Electronic Letters. 2006;**42**(6):370–371

[5] Lee S, Randel S, Breyer F, and Koonen A. PAM-DMT for intensity-modulated and direct-detection optical communications. IEEE Phtonics Technology Letters. 2009;**21**(23):1749–1751

[6] Tsonev D, Sinanovic S, and Haas H. Novel unipolar orthogonal frequency division multiplexing (U–OFDM) for optical wireless. In: Proc. IEEE Veh. Tech. Conf.; May 2012; Yohohama, Japan. pp. 1–5

[7] Fernando N, Hong Y, and Viterbo E. Flip-OFDM for unipolar communication systems. IEEE Transcations on Communications. 2012;**60**(12):3726–3733

[8] Dissanayake S and Armstrong J. Comparison of ACO-OFDM, DCO-OFDM and ADO-OFDM in IM/DD systems. Journal of Lightwave. Technology. 2013;**31**(7):1063–1172

[9] Wang Q, Qian C, Guo X, Wang Z, Cunningham D, and White I. Layered ACO-OFDM for intensity-modulated direct-detection optical wireless transmission. Optics Express. 2015;**23**(9):12382–12393

[10] Ranjha B and Kavehrad M. Hybrid asymmetrically clipped OFDM-based IM/DD optical wireless system. Journal of Optical Communications and Networking. 2014;**6**(4):387–396

[11] Tsonev D and Haas H. Avoiding spectral efficiency loss in unipolar OFDM for optical wireless communication. In: Proc. IEEE ICC; Jun. 2014; Sydney, Australia. pp. 3336–3341

[12] Wen M, Cheng X, and Yang L. Index Modulation for 5G Wireless Communications. Springer; 2017. 154 p

[13] Basar E. Index modulation techniques for 5G wireless networks. IEEE Communications Magazine. 2016;**54**(7):168–175

[14] Frenger P and Svensson N. Parallel combinatory OFDM signalling. IEEE Transactions on Communications. 1999;**47**(4):558–567

[15] Abu-Alhiga R and Haas H. Subcarrier-index modulation OFDM. In: Proc. 20th IEEE Int. Symp. Indoor Mobile Radio Commun.; Sep. 2009; Tokyo, Japan. pp. 177–181

[16] Tsonev D, Sinanovic S, and Haas H. Enhanced subcarrier index modulation (SIM) OFDM. In: Proc. IEEE GLOBECOM Workshops; Dec. 2011; TX, USA. pp. 728–732

[17] Basar E, Aygolu U, Panayirci E, and Poor H. Orthogonal frequency division multiplexing with index modulation. IEEE Transactions on Signal Processing. 2013;**61**(22):5536–5549

[18] Ishikawa N, Sugiura S, and Hanzo L. Subcarrier-index modulation aided OFDM-Will it work?. IEEE Access. 2016;**4**:2580–2593

[19] Xiao Y, Wang S, Dan L, Lei X, Yang P, and Xiang W. OFDM with interleaved subcarrier-index modulation. IEEE Communication Letters. 2014;**18**(8):1447–1450

[20] Basar E. OFDM with index modulation using coordinate interleaving. IEEE Wireless Communications Letters. 2015;**4**(4):381–384

[21] Fan R, Yu Y, and Guan Y. Generalization of orthogonal frequency division multiplexing with index modulation. IEEE Transactions on Wireless Communication. 2015;**14** (10):5350–5359

[22] Zheng B, Chen F, Wen M, Ji F, Yu H, and Liu Y. Low-complexity ML detector and performance analysis for OFDM with in-phase/quadrature index modulation. IEEE Communication Letters. 2015;**19**(11):1893–1896

[23] Yang X, Zhang Z, Fu P, and Zhang J. Spectrum-efficient index modulation with improved constellation mapping. In: Proc. IEEE HMWC; Oct. 2015; Xi'an, China. pp. 91–95

[24] Basar E. Multiple-input multiple-output OFDM with index modulation. IEEE Signal Processing Letters. 2015;**22**(12):2259–2263

[25] Basar E. On multiple-input multiple-output OFDM with index modulation for next generation wireless networks. IEEE Transactions on Signal Processing. 2016;**64**(15):3868–3878

[26] Wen M, Cheng X, Yang L, Li Y, Cheng X, and Ji F. Index modulated OFDM for underwater acoustic communications. IEEE Communications Magazine. 2016;**54**(5):132–137

[27] Wen M, Li Y, Cheng X, and Yang L. Index modulated OFDM with ICI self-cancellation in underwater acoustic communications. In: Proc. IEEE Asilomar Conf. Signals, Syst, Comput.; Nov. 2014; Pacific Grove, CA, USA. pp. 338–342

[28] Cheng X, Wen M, Yang L, and Li Y. Index modulated OFDM with interleaved grouping for V2X communications. In: Proc. IEEE Int. Conf. Intell. Transp. Syst.; Oct. 2014; Qingdao, China. pp. 1097–1104

[29] Mao T, Wang Z, Wang Q, Chen S, Hanzo L. Dual-mode index modulation aided OFDM. IEEE Access. 2017;**5**:50–60

[30] , Mao T, Wang Q, and Wang Z. Generalized dual-mode index modulation aided OFDM. IEEE Communication Letters. 2017;**21**(4):761–764

[31] Wen M, Cheng X, and Yang L. Optimizing the energy efficiency of OFDM with index modulation. In: Proc. IEEE Int. Conf. Commun. Syst.; Nov. 2014; Macau, China. pp. 31–35

[32] Li W, Zhao H, Zhang C, Zhao L, and Wang R. Generalized selecting sub-carrier modulation scheme in OFDM system. In: Proc. IEEE ICC Workshops; Jun. 2014; Sydney, NSW, Australia. pp. 907–911

[33] Basar E, Panayirci E. Optical OFDM with index modulation for visible light communications. In: Proc. IEEE Int. Workshops on Opt. Wirel. Commun.; Sep. 2015; Istanbul, Turkey. pp. 11–15

[34] Mao T, Jiang R, and Bai R. Optical dual-mode index modulation aided OFDM for visible light communications. Optics Communications. 2017;**391**:37–41

[35] Wang Z, Wang Q, Chen S, and Hanzo L. An adaptive scaling and biasing scheme for OFDM-based visible light communication systems. Optics Express. 2014;**22**(10):12707–12715

[36] Wang Q, Wang Z, and Dai L. Asymmetrical hybrid optical OFDM for visible light communications with dimming control. IEEE Photonics Technology Letters. 2015;**27**(9):974–977

[37] Wang Q, Wang Z, Dai L, and Quan J. Dimmable visible light communications based on multilayer ACO-OFDM. IEEE Photonics Journal. 2016;**8**(3)

[38] Wang Q, Wang Z, and Dai L. Multiuser MIMO-OFDM for visible light communications. IEEE Photonics Journal. 2015;**7**(6)

[39] Basar E, Panayirci E, Uysal M, and Haas H. Generalized LED index modulation optical OFDM for MIMO visible light communications. In: IEEE ICC 2016; May 2016; Kuala Lumpur, Malaysia

[40] Sugiura S, Chen S, and Hanzo L. Generalized space-time shift keying designed for flexible diversity-, multiplexing- and complexity-tradeoffs. IEEE Transactions on Wireless Communications. 2011;**10**(4):1144–1153

Chapter 5

The Novel PAPR Reduction Schemes for O-OFDM-Based Visible Light Communications

Tian Zhang, Jun Yao and Shuxu Guo

Additional information is available at the end of the chapter

http://dx.doi.org/10.5772/intechopen.68763

Abstract

In this chapter, we propose two novel peak-to-average power ratio (PAPR) reduction schemes for the asymmetrically clipped optical orthogonal frequency division multiplexing (ACO-OFDM) scheme used in the visible light communications (VLC) system. In the first scheme, we implement the Toeplitz matrix based Gaussian blur method to reduce the high PAPR of ACO-OFDM at the transmitter and use the orthogonal matching pursuit algorithm to recover the original ACO-OFDM frame at the receiver. Simulation results show that for the 256-subcarrier ACO-OFDM system a ~6 dB improvement in PAPR is achieved compared with the original ACO-OFDM in terms of the complementary cumulative distribution function (CCDF), while maintaining a competitive bit-error rate performance compared with the ideal ACO-OFDM lower bound. In the second scheme, we propose an improved hybrid optical orthogonal frequency division multiplexing (O-OFDM) and pulse-width modulation (PWM) scheme to reduce the PAPR for ACO-OFDM. The bipolar O-OFDM signal without negative clipping is converted into a PWM format where the leading and trailing edges carry the frame synchronization and modulated information, respectively. The simulation and experimental results demonstrate that the proposed OFDM-PWM scheme offers a significant PAPR reduction compared to the ACO-OFDM with an improved bit error rate.

Keywords: visible light communications, peak-to-average power ratio, optical orthogonal frequency division multiplexing, Gaussian blur, pulse-width modulation

1. Introduction

The rapid development of solid state lighting technologies has made the visible light communications (VLCs) a promising complementary scheme in the widely used radio frequency

(RF)-based wireless communications in certain indoor and possible outdoor applications [1]. The main challenge of the VLC technology is the lack of sufficiently large bandwidth for modulation due to the fact that the white light emitting diodes (LEDs) used for VLC usually have a very small bandwidth, blue LED (BLED). Therefore, the transmission capacity of the VLC is limited [2]. In order to address this issue and make good use of the limited BLED with the aim of increasing the transmission throughput, a number of schemes including (i) blue filtering at the receiver (Rx) to remove the slow phosphor part of the spectrum, which increases BLED to ~20 MHz but at the cost of high power loss, (ii) high spectrally efficient modulation schemes, such as a variant of optical orthogonal frequency division multiplexing (O-OFDM), discrete multi-tone modulation (DMT), and multi-band carrier-less amplitude and phase modulation, have been extensively investigated in the literature [3].

In intensity modulation and direct detection (IM/DD)-based VLC systems, the traditional complex and bipolar OFDM is modified to a real and unipolar format [4]. Though OFDM offers a number of advantages such as efficient use of the spectrum, there are many issues in OFDM-based VLC systems including (i) a high peak-to-average power ratio (PAPR), (ii) performance degradation due to non-linear power-current (P-I) characteristics of LEDs, (iii) limited support for dimming, and (iv) reduced throughput due to the use of longer cyclic prefix as a result of longer tails of the impulse response. Among these, the high PAPR is by far the most detrimental to the performance of OFDM-based VLC systems [5]. A higher PAPR would lead to more severe distortion and clipping because of the non-linear P-I characteristics of the LED [6]. This leads to a decreased signal-to-quantization noise ratio (SQNR) in both analog-to-digital (A/D) and digital-to-analog (D/A) converters while not fully utilizing the wide dynamic range of LEDs [7].

A number of techniques to mitigate the high PAPR requirement in O-OFDM have been reported in the literature including amplitude clipping [8], trellis coding, which reduces the average optical power [9], and block coding, which maps the vector of k-information bits to be transmitted and the vector of symbol amplitudes modulated onto the N-subcarrier [10] at the cost of increased transmission bandwidth. Signal transformation based on selected mapping (SLM) has also been used to reduce the signal peak values in O-OFDM systems [11]. In Ref. [12], the signal peak values are reduced by applying a pilot symbol (PS) phase rotation technique to the original OFDM signal. In Ref. [13], lower-order modulation on subcarriers, which suffers the most distortion, is used to achieve a high throughput for OFDM-VLC at a sampling rate of six times the available system bandwidth with reduced PAPR. The phase of PS is chosen based on the SLM algorithm while the maximum likelihood criterion is used at the Rx to estimate the PS. These techniques achieve PAPR reduction at the expense of increasing the transmit signal power, bit error rate (BER), data rate loss, computational complexity, and so on.

In this chapter, we introduce two novel PAPR reduction techniques for multicarrier transmission-based VLC system with some simulation and experiment demonstrations. The first scheme is named as Gaussian blur (GB) [14], which is enlightened by the blur operation used in image processing. The two-dimensional GB has been widely implemented in graphics software to reduce image noise and details by convolution operations [15]. Similar to this idea of blurring images, we apply the one-dimensional GB to the time-domain signal of asymmetrically clipped optical orthogonal frequency division multiplexing (ACO-OFDM) system for reducing the high PAPR in VLC. Simulation results show that for the 256-subcarrier ACO-OFDM system, a ~6 dB improvement in PAPR is achieved in terms of the complementary cumulative distribution function (CCDF). However, the PAPR problem is only alleviated temporarily but not eliminated. In addition, considering that the indoor environment (i.e., the channel) is relatively static with no deep fading, the multipath-induced interference is not a major issue as is the case in an outdoor environment for RF-based wireless systems, and the use of O-OFDM needs additional operation to mitigate the high PAPR. Therefore, there is an open question: are there any advantages in transmitting the OFDM signal over a VLC channel or does it need to be converted into a digital format to avoid all the issues outlined above. To address this question, we further propose the second PAPR reduction scheme, which is named as the hybrid OFDM-PWM modulation [16], for converting the OFDM signal into a pulse-width modulation (PWM) format prior to IM of the LED in order to mitigate the high PAPR. The high PAPR in O-OFDM is no longer a major issue in OFDM-PWM as LEDs are only switched between "on" and "off". A similar technique is proposed in [17] where a linear mapping function is used to convert OFDM samples into PWM. However, the required bandwidth of proposed scheme in [17] exceeds the O-OFDM scheme, thus leading to a bit error rate (BER) penalty. In our improved OFDM-PWM scheme, we take advantage of the anti-symmetry property of the time-domain ACO-OFDM signal and hence convert only the first half of samples of the ACO-OFDM frame and extend the pulse width of PWM by a factor Nc. This ensures that the proposed OFDM-PWM has the same bandwidth requirement as the ACO-OFDM. Simulation and experimental results demonstrate that our proposed OPDM-PWM scheme has an improved BER performance compared with the original ACO-OFDM. Furthermore, the advantages of the scheme in Ref. [17], such as reduced PAPR, resilience to LED non-linearity, and higher luminance level, are maintained.

2. The PAPR of O-OFDM and the CCDF of PAPR

OFDM is widely adopted in RF and optical communications including free space optics and VLC due to its huge data transmission capability, high spectral efficiency, and resilience to the channel-induced impairments. Although the OFDM modulated waveform has many advantages in both the RF and optical domain, its predominant drawback is that the signal profile has intermittent peaks that occur throughout the length of the OFDM signal contributing immensely to the peak-to-average power ratio (PAPR). With non-constant amplitude signals, it is important to improve the PAPR of those signals. The presence of these high peaks means that the optical source will have to operate outside its linear region to accommodate the full amplitude signal swings. This is very undesirable as it increases the level of distortion present in the transmitted signal, which results in poor system performance. Here, the definition of PAPR for a discrete ACO-OFDM signal $x(n)$ is given to evaluate the ratio of the maximum instantaneous power to the average power. To better approximate the PAPR of $x(n)$, the

O-OFDM signal samples $x(n)$ are obtained by oversampling L times. In [11], it is demonstrated that a fourfold oversampling factor ($L = 4$) is enough to provide an accurate measure of the PAPR value. Thus, the electrical PAPR of a single symbol O-OFDM signal is given as:

$$PAPR \triangleq \frac{\max|x(n)|^2}{E\left[|x(n)|^2\right]} \quad \text{for } 0 \leq n \leq N \times L{-}1 \tag{1}$$

where $E[.]$ denotes the statistical expectation, n is the number of samples for each OFDM frame in the time domain, and $x(n)$ is real and unipolar.

The cumulative distribution function (CDF) of the PAPR is one of the most frequently used performance measures for PAPR reduction techniques. In the literature, the complementary CDF (CCDF) is commonly used instead of the CDF itself. The CCDF of the PAPR denotes the probability that the PAPR of a data block exceeds a given threshold PAPR_0 as given by [18]:

$$\text{CCDF} = P(\text{PAPR} > \text{PAPR}_0) = 1{-}P(\text{PAPR} \leq \text{PAPR}_0) = 1{-}\text{CDF} \tag{2}$$

3. GB-based PAPR reduction scheme

3.1. System overview

The block diagram of the GB-based ACO-OFDM system is depicted in **Figure 1**. In ACO-OFDM systems, the time-domain signal needs to be both real and unipolar. The bipolar signal at the output of Inverse Fast Fourier Transform (IFFT) is converted into unipolar by clipping the negative value to zero. Unlike the conventional ACO-OFDM system, in our scheme, the clipped output of IFFT is fed into theGaussian blur (GB) module to generate the alternative output sequence $S_i(n)$. Then, the sequence $S(n)$ with the lowest PAPR is applied to the optical driver module.

At the receiver side, the received signal is influenced by the noise sources in a real scenario. The dominant noise source in an indoor wireless optical channel is the ambient light-induced shot noise [19], which is modeled as the additive white Gaussian noise (AWGN). Thus, the received signal is given by:

$$y(n) = \eta \cdot \; S(n) \otimes h + Z(n), \;\; S(n) = \phi \cdot x(n) \tag{3}$$

where η is the photoelectric conversion efficiency, ϕ is a standard Toeplitz matrix, $\phi{\cdot}x(n)$ is the matrix representation of the Gaussian blur operation, $x(n)$ is the original ACO-OFDM signal, $S(n)$ is the alternative output signal with the lower PAPR, h is the channel response, $Z(n)$ is AWGN with zero mean and variance of δ_z^2, and $\otimes$ denotes the convolution operation.

We assume that the channel state information is perfectly known in advance. According to the transmitter design of ACO-OFDM, a half of the transmitted signals are clipped to be zeros. Therefore, the time-domain ACO-OFDM signal is a standard sparse signal adopted in the compressed sensing (CS) theory. Considering the special time-domain structure of ACO-OFDM symbols with the system complexity, we choose the orthogonal matching pursuit

http://dx.doi.org/10.5772/intechopen.68763

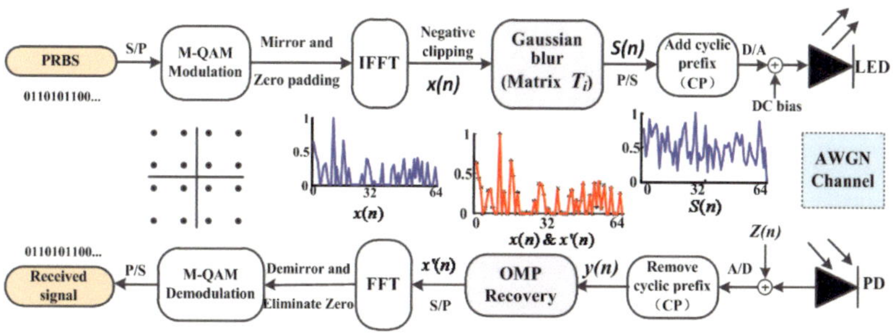

Figure 1. A block diagram of GB-based ACO-OFDM system is shown. S/P: serial-to-parallel converter, P/S: parallel-to-serial converter, and PD: photodetector.

(OMP) approach [20], which is widely used in sparse signal reconstruction in our simulation for recovering the original ACO-OFDM signal. Following the fast Fourier transformation (FFT) and demodulation process, we regenerate the original binary data stream.

3.2. GB algorithm

GB is a widely used image processing algorithm, which has the advantages of being very "smooth" and also circularly symmetric, so that edges and lines in various directions are treated similarly. It is well known that the two-dimensional GB operations are employed as the convolution in image processing to blur the images. The two-dimensional Gaussian kernel is defined as:

$$G(x, y) = \frac{1}{2\pi\delta^2} \cdot e^{\frac{-(x^2+y^2)}{2\delta^2}} \tag{4}$$

Applying GB to the original image $I(x, y)$ by convolving with Gaussian kernel function $G(x, y)$ to realize the blur operations, we have:

$$\begin{aligned} I_g(x, y) &= I(x, y) \otimes G(x, y) \\ &= \sum_{m,n} I(m, n) \cdot G(x-m, y-n) \end{aligned} \tag{5}$$

Similar to image processing, we apply the GB operation to the one-dimensional ACO-OFDM signal in optical communications. One-dimensional Gaussian kernel function $G(n)$ is implemented to convolve with the one-dimensional ACO-OFDM signal $x(n)$ as given by:

$$S(n) = \sum_{k=1}^{N} G(n-k)x(k), n = 1, 2 \ldots M \tag{6}$$

where N and M are the lengths of the ACO-OFDM signal and the alternative output $S(n)$, respectively.

Expanding Eq. (6), we have:

$$\begin{pmatrix} g(0) & g(-1) & \dots g(-N+1) \\ g(1) & g(0)\dots & g(-N+2) \\ \vdots & \vdots & \\ g(M{-}1) & g(M{-}2) & \dots g(M{-}N) \end{pmatrix} \cdot \begin{pmatrix} x_1 \\ x_2 \\ \vdots \\ x_N \end{pmatrix} = \begin{pmatrix} S_1 \\ S_2 \\ \vdots \\ S_M \end{pmatrix} \tag{7}$$

Corresponding to the CS theory, matrix $g(.)$ is the measurement matrix ϕ, matrix $x(.)$ is the sparse ACO-OFDM signal, and matrix $S(.)$ is the significant measurement, which has a lower PAPR compared with the original $x(.)$. From Eq. (7), we can see that the g(.) matrix is a standard Toeplitz matrix and its elements are Gaussian random variables with zero mean and variance of δ_z^2. Therefore, only (M + N–1) elements are used to construct G, which means that only $\log_2(M+N{-}1)$ bits are needed to be transmitted as the side information [21]. In this module, a "blur" output sequence $S(n)$ is generated by taking the weighted sum of all the elements in $x(n)$ using the Gaussian distribution coefficients sequence $g(n)$ as the weights. Due to the random characteristic of the Gaussian function generated by the Matlab software which is used in our study, the generated output $S(n)$ is not the same as each other. Here, we construct a series of Toeplitz matrices T_i to obtain the lowest PAPR signal from $S_i(n)$, where i is the number of candidate matrices. Note that we only select the output signal $S_i(n)$ with the lowest PAPR for transmission.

Similar to the SLM method, the performance of the proposed scheme depends on the value of i and the elements of T. However, lower PAPR can be potentially attained for all ACO-OFDM signals as shown in **Table 1**.

To trade-off PAPR reduction and the system complexity, we have used $i = 6$ in our simulations to achieve ~6 dB improvement.

3.3. OMP recovery algorithm

The CS technique has been widely applied in signal reconstruction, medical imaging, radar, remote sensing, and other signal processing fields [22]. The main advantage of this theory is that it can recover original signals or images from far fewer samples or measurements. To realize this, CS has two fundamental criteria: (i) the original signal is sparse or compressible and (ii) the measurement matrix ϕ satisfies the restricted isometry property (RIP).

Suppose the transmission signal $x \in R^N$ is an N-dimensional sparse signal with sparsity K, $K \ll N$. The sparse reconstruction problem of CS is defined as recovering the sparse signal from the observed vector of measurements $S \in R^M$. The classical mathematical expression of CS measurement is given as:

$$S = \phi x + z \tag{8}$$

where $\phi \in R^{(M\times N)}$ is a known measurement matrix with $M \ll N$ and $z \in R^M$ denotes the measurement noise and model error. Note that measurement matrix ϕ needs to satisfy the RIP:

$$(1{-}\delta_k)\|x\|_{l_2}^2 \le \|\phi x\|_{l_2}^2 \le (1+\delta_k)\|x\|_{l_2}^2 \tag{9}$$

i_{th} PAPR	1	2	3	4	5	6	7	Original
x_1 (dB)	9.0	9.1	**8.0**	10.9	10.7	9.1	9.3	11.6
x_2 (dB)	9.3	9.5	8.4	8.4	9.0	**7.7**	11	12.4
x_3 (dB)	9.0	8.6	**7.5**	9.1	8.2	8.5	10.4	12.8

Note the simulation results obtained from three random 16-QAM ACO-OFDM signals (256-subcarrier). The numbers in bold are the minimum PAPR from a set of *i*- measurements.

Table 1. PAPR reduction results for original ACO-OFDM with GB.

where δ_k is a restricted isometric constant. For more details, please see [23]. Then, by resolving Eq. (8) with corresponding sparse signal reconstruction algorithm, one can obtain the estimation $\tilde{x}$. Two popular reconstruction algorithms are the convex optimization algorithm and the iterative greedy pursuit algorithm.

For the ACO-OFDM system, we can see that the asymmetrically clipped output of N-point IFFT is a sparse signal in the time domain, and the Toeplitz matrix composed of Gaussian random variables meets the restricted isometry property (RIP) standard as adopted in [23]. Therefore, the OMP algorithm, which has the advantages of lower computation complexity, rapid recovery speed, and higher reconstruction accuracy compared to other algorithms, can be employed in our scheme [24]. The equivalent electrical voltage signal $\tilde{Y}$ can be employed to reconstruct $\tilde{x}$ as outlined below:

$$\min\|\tilde{x}\|_1 \; s.t. \; T\tilde{x} = \tilde{Y} \tag{10}$$

where $T\in R^{(M\times N)}$ is the observation matrix ϕ and $\tilde{Y}\in R^M$ denotes the measured values of S. The OMP algorithm used in this chapter is summarized as below:

Step 1. Setting the residual value $r_0 = \tilde{Y}$, pre-recovery $\tilde{x} = 0$, and iterations number $n = 1.4*K$ ($n\geq K$), where K is the sparsity of x.

Step 2. Finding the max<T(:;col), r_0>and its position. Note that $M\geq C*K*\log_2(N/K)$, where C is a constant, depends on each instance.

Step 3. Calculating the reconstructed $\tilde{x}$ by min $\|\tilde{Y}-T\tilde{x}\|_2$ and updating the residual $r_n = \tilde{Y}-\frac{<T(\text{col}),\tilde{Y}>}{<T(\text{col}),T(\text{col})>}T(\text{col})$ by the least square algorithm.

Step 4. If the criterion has not been satisfied, then return to step 2.

3.4. Simulations and results

Simulations are conducted to compare the most widely used amplitude clipping method, classical SLM technique and the GB method, and the key parameters adopted are shown in **Table 2**.

Figure 2 shows the CCDF against the threshold $PAPR_0$ for the unmodified ACO-OFDM ($N = 256$) with the classical SLM, amplitude clipping (CL of 0.8), and the proposed method. Also shown is

Method	Subcarrier	Modulation	Parameter	Length
Clipping	256	16-QAM	CL = 0.8	—
SLM	256	16-QAM	$i = 6$	$N/4$
GB	256	16-QAM	$i = 6$	$M + N - 1$

Table 2. Simulation parameters for PAPR reduction schemes.

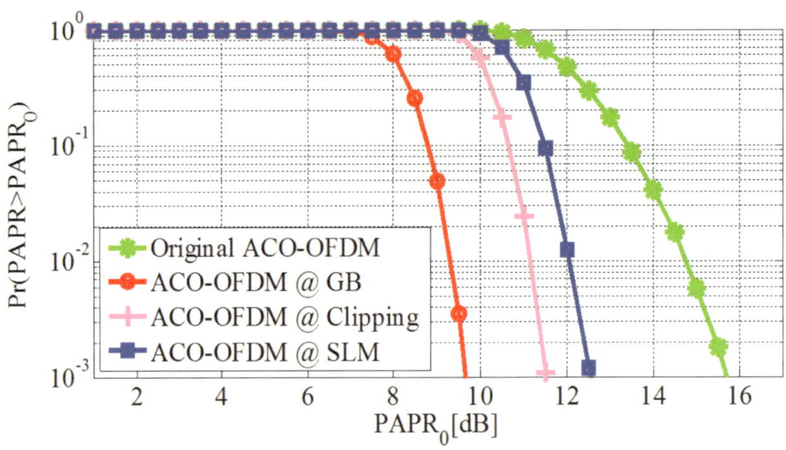

Figure 2. PAPR analysis for ACO-OFDM signals ($N = 256$) with SLM, clipping (CL = 0.8), and GB techniques ($i = 6$).

the plot for the original ACO-OFDM with no PAPR reduction scheme. The horizontal and vertical axes represent the threshold for the PAPR and the probability that the PAPR of a data block exceeds the threshold, respectively. It is shown that at CCDF of 10^{-3} the PAPR requirements are < 10, ~11.5, and > 12 dB for ACO-OFDMs with GB, amplitude clipping, and SLM, respectively, compared with the < 16 dB for the original ACO-OFDM. From the comparison, we can conclude that the proposed method offers improved PAPR reduction. Note that the simplest amplitude clipping method can achieve a much significant PAPR reduction at the cost of the unrecoverable performance degradation.

Next, we want to recover the original ACO-OFDM signal $x(n)$ from $\tilde{Y}$ at the receiver. For perfect reconstruction of the original signal, the row size of T must meet the following condition as outlined in [24]:

$$M \geq C \cdot K \cdot \log_2(N/K) \tag{11}$$

In ACO-OFDM, K is $N/2$. Therefore, M should be larger than the product of C and K to meet the signal reconstruction requirements. The signal reconstruction accuracy is evaluated by the reconstruct error (RE), which is given by:

$$\text{RE} = \frac{\|x(n)' - x(n)\|_2}{\|x(n)\|_2} \tag{12}$$

where $x(n)'$ is the reconstructed ACO-OFDM signal and $\|x(n)\|_2$ denotes the second moment norm, $\|x(n)\|_2 = \sqrt[2]{|x_1|^2 + |x_2|^2 + \cdots + |x_n|^2}, n = 1, 2, \ldots, N$.

Following a number of simulations with a range of C values shown in **Table 3**, we found that RE is reduced significantly with C increased from 1 to 2. However, for $C > 2$, the improvement in RE is very small, and $S(n)$ is longer than the original signal $x(n)$ measured in the time domain, since $M > N$. For the trade-off between RE and the data rate, we have selected $C = 2$ as the optimal value for the proposed system. Therefore, the Toeplitz matrix is a square matrix $T_{N\times N}$ with no incurred bandwidth penalty.

Figure 3 shows the reconstruction results for 256-subcarrier-based ACO-OFDM signal for an SNR of 30 dB using the OMP recovery algorithm. Note that in typical indoor VLC systems, the most widely used SNR is 30 dB [19]. We observe the faithful reconstruction of the distortion-less original signal for an SNR of 30 dB compared to the original ACO-OFDM signal. We have used $1 \times e^3$ random ACO-OFDM signals to calculate the RE over an SNR range of 2030 dB as shown in the inset in **Figure 3**. Note that the improvement in RE is < 0.06, which is far below the acceptable value of < 0.1 for the entire system performance.

Figure 4 illustrates the BER performance against the SNR for the uncoded 256-subcarrier ACO-OFDM with amplitude clipping and the proposed scheme. In practice, higher peak-to-average power ratio (PAPR) can lead to non-linear distortions and clipping in systems with peak-power limitations ($PAPR_0$), thus leading to degradation of the system BER performance. In this chapter, we define the ideal ACO-OFDM as having an infinite $PAPR_0$ and, therefore, the adverse effect of higher PAPR of the original ACO-OFDM signal is not considered in the BER

C	1.0	1.5	2.0	2.5	3.0	3.5	4.0
RE	14.25	0.43	0.031	0.022	0.021	0.020	0.017
Data rate	200%	133%	100%	80%	67%	57%	50%

Note that the simulation results are obtained for 256-subcarrier ACO-OFDM with an SNR of 30 dB.

Table 3. RE values and data rate against different C values.

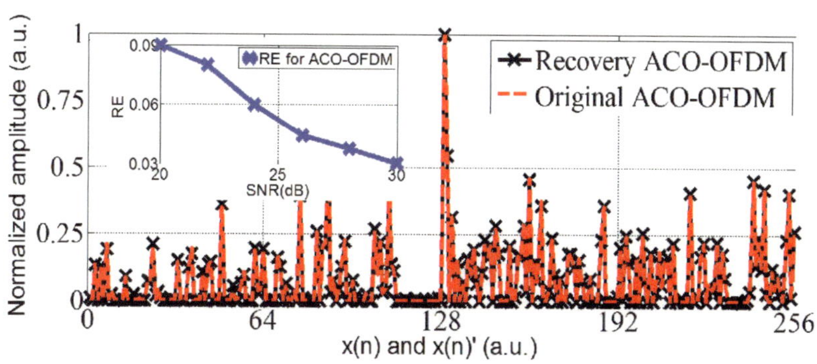

Figure 3. RE analysis for ACO-OFDM (with $N = 256$) for SNR of 30 dB. The inset shows the RE for 20 < SNR < 30 dB.

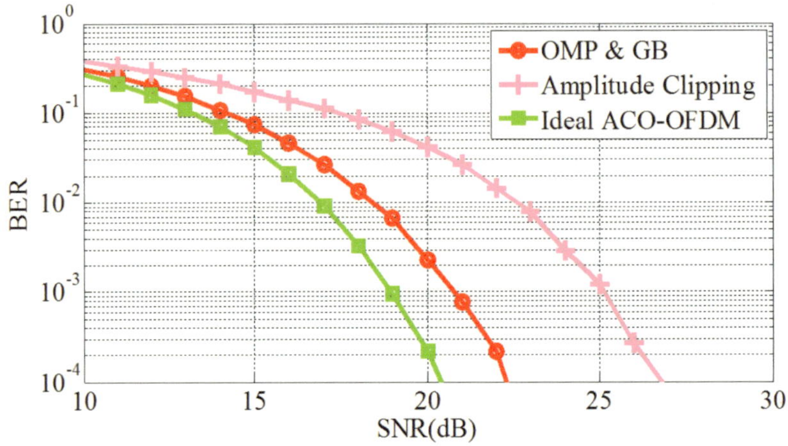

Figure 4. BER performance against the SNR for 256-subcarrier ACO-OFDM systems with amplitude clipping (CL = 0.8) and the proposed method ($i = 6$).

analysis. The BER performance of an ideal ACO-OFDM is used as a reference lower bound as in reference [25]. Thus, the recovery accuracy of the OMP algorithm is indirectly illustrated by the BER performance. At a BER of 10^{-4}, the proposed scheme requires only a ~2 dB of additional SNR compared to the ideal ACO-OFDM scheme. For ACO-OFDM with amplitude clipping, the SNR penalty is almost 5 dB compared to the proposed scheme. It also can be seen that when the SNR is larger than 20 dB, the BER of the proposed scheme is within the bound of the forward error correction (FEC) limits. And as for the VLC, these SNR values can be obtained by the need for illumination. From the results shown in **Figures 2–4**, we conclude that the proposed scheme can achieve a significant reduction in PAPR with a competitive recovery performance.

4. OFDM-PWM-based PAPR reduction scheme

4.1. System overview

The block diagram of the proposed OFDM-PWM scheme is shown in **Figure 5**, where the bipolar discrete-time O-OFDM samples $x(n)$ are first generated using the standard ACO-OFDM process as outlined in Ref. [26] but with no zero clipping operation prior to conversion to PWM. In the traditional O-OFDM such as DC-biased optical OFDM (DCO-OFDM), a DC bias is added to the signal before clipping the negative residual signals to ensure positive signal amplitude. However, due to a high PAPR, the DC bias required is very high, and it is also necessary to clip the DC-biased signal to clip the negative residue. ACO-OFDM overcomes the high DC-biased requirements by encoding information only on odd harmonics and clipping the negative signals at zero, which do not lead to any information loss [26]. Due to the Hermitian symmetry requirements to ensure a real-time-domain signal and with only odd subcarriers being used, only one-fourth of the subcarrier is actually encoded with the data. As a result, the spectral efficiency of ACO-OFDM is half of that of DCO-OFDM [27]. Therefore, the

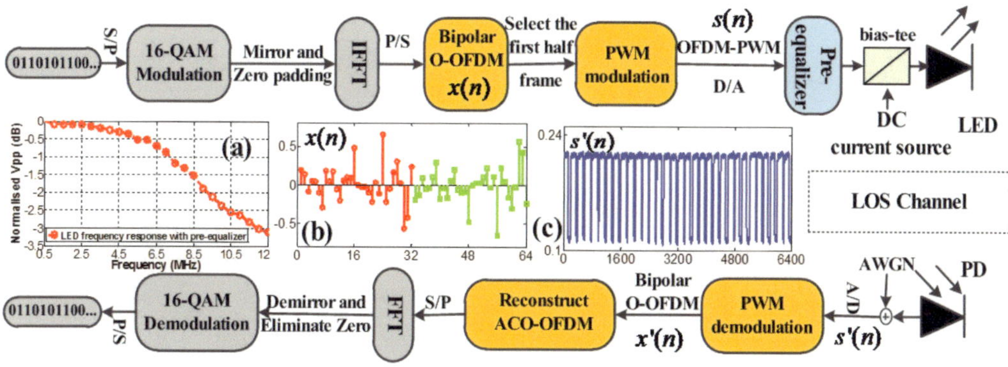

Figure 5. A block diagram of OFDM-PWM (a) LED frequency response with pre-equalizer, (b) bipolar O-OFDM before asymmetrical clipping, and (c) received OFDM-PWM waveform. LOS: line of sight.

resulting time-domain signal is a real-valued signal referred to as the O-OFDM frame with the anti-symmetry property as defined by Armstrong and Schmidt [27] and as shown in **Figure 5 (b)**:

$$x_k = -x_{k+N/2},\ 0 < k < N/2 \tag{13}$$

where N is the number of IFFT points.

In the OFDM-PWM scheme, since PWM can be used to linearly represent the bipolar signal, the zero clipping operation in ACO-OFDM is no longer necessary. Due to ACO-OFDM's anti-asymmetric time-domain characteristics, the first half of the bipolar samples (with no zero clipping) is sufficient to represent the entire unipolar ACO-OFDM. Hence, instead of converting the entire O-OFDM symbol into PWM, converting only the first $N/2$ samples would be sufficient, which does not lead to any loss of information. Thus, we discard the unnecessary last half of samples and save half of the time slots for further compensating the high bandwidth requirement for PWM, see **Figure 6(b)** (Digital OFDM-PWM). The generated OFDM-PWM signal is used for IM of the LED. At the receiver (Rx), the OFDM signal is extracted from OFDM-PWM, and the standard OFDM demodulation process is thereafter employed, which is the reverse of the transmitter (Tx) process, as outlined in Ref. [27].

PWM can be generated in the analog or digital domain. In the analogue PWM, the input signal (i.e., $x(n)$ in this case) is compared with a reference carrier signal (i.e., a ramp waveform) of frequency f_c as shown in **Figure 6(a)**. Note that the Nyquist condition $f_c \geq 2NT_s^{-1}$needs to be met, where T_s is the OFDM symbol period. In the digital PWM, the digital version of $x(n)$ is compared with a counter, which rests every 2^L where L is the bit resolution adopted in this chapter. In the PWM scheme, the falling edge carries the information whereas the rising edge is used for synchronization (more information on PWM can be found in Ref. [28]). Thus, the converted PWM signal shown in **Figure 6(a)** is given by:

$$PWM(t) = \begin{cases} C, & 0 \leq t \leq \tau \\ 0, & \tau < t \leq T_c \end{cases} \tag{14}$$

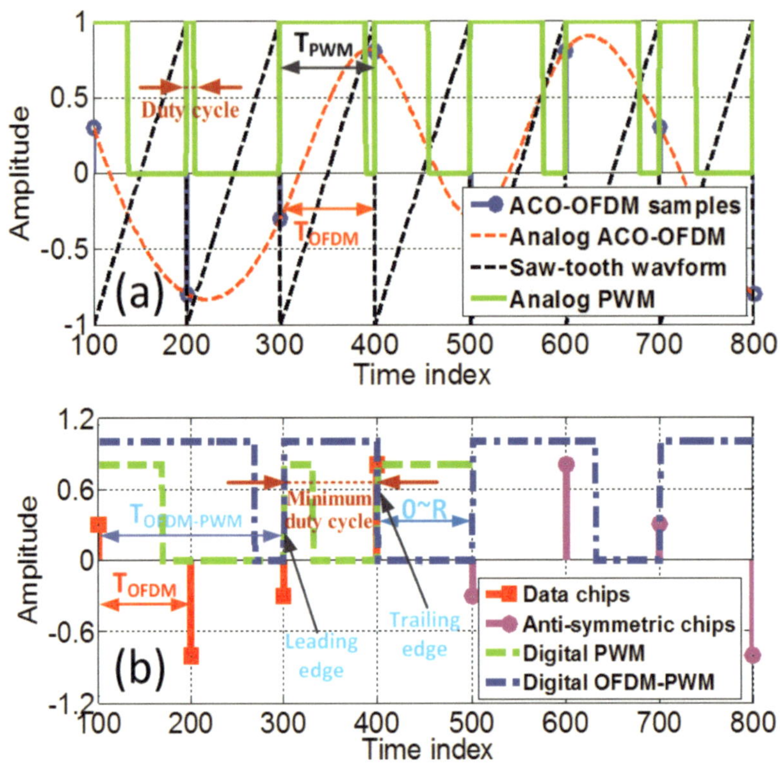

Figure 6. Time-domain waveforms for (a) ACO-OFDM, ramp, and PWM and (b) the process of the digital OFDM-PWM generation.

where T_c is the OFDM symbol sampling period, C is a constant corresponding to the peak amplitude of O-OFDM, and τ, given by Eq. (15), is the modulated pulse width that changes linearly with the instantaneous value of $x(n)$ as given by:

$$\tau(n) = \frac{x(n) - x_{\min}}{K} \tag{15}$$

where K is the PWM carrier signal slope factor or the modulation index defined as:

$$K = \frac{x_{\max} - x_{\min}}{T_c} \tag{16}$$

Note that $x_{\max}$ and $x_{\min}$ are the maximum and minimum amplitudes of O-OFDM symbols, respectively. In order to simplify generation of the digital PWM signal, the sampling period is divided into R discrete intervals with $R \geq M$, where M is quadrature amplitude modulation (QAM) modulation order. Therefore, Eqs. (15 and 16) are rewritten in a discrete form as:

$$K = \frac{x_{\max} - x_{\min}}{R} = \frac{(x(n) - x_{\min})}{N_\tau}, \quad n = 1, \ldots, N/2 \tag{17}$$

where N_τ is the corresponding discrete interval of $\tau(n)$.

In the converted PWM scheme, see **Figure 6(a)**, the minimum pulse duration $\tau(n)_{\min} = 1/R$ is significantly smaller than T_c. Hence, the bandwidth requirement for PWM is R times higher than ACO-OFDM. In order to overcome the high bandwidth requirement, the duty cycle of PWM signal is further extended by a factor of N_c, where $0 \le N_c \le R$ as shown in **Figure 6(b)** with blue lines. The value of N_c can be adjusted either to obtain a maximum throughput by setting $N_c = 0$ or a required dimming level by setting $N_c > 0$. In this study, we have increased the minimum pulse width to R by setting $N_c = R$ (see **Figure 6(b)**). This ensures that the PWM pulse duration is at least equal to or greater than T_c, which means that the proposed OFDM-PWM scheme has the same bandwidth requirement as that of ACO-OFDM. In addition, the increased time factor N_c is equal to the OFDM sample period, which indicates that we have taken full advantage of the last half time slots saved from the previous sampling part. Thus, the transformation from ACO-OFDM to OFDM-PWM is a lossless process. The proposed OFDM-PWM signal $s(n)$ is given by:

$$s(n) = \begin{cases} C, & 0 \le N_{s(n)} \le (N_\tau + N_c) \\ 0, & (N_\tau + N_c) < N_{s(n)} \le 2R \end{cases}, n = 1, \ldots, N/2 \tag{18}$$

where $N_s(n)$ is the total number of discrete intervals corresponding to the duty cycle of each OFDM-PWM signal.

Following the optical to electrical (O-E) conversion at the Rx, the received OFDM-PWM signal $s'(n)$ is given by:

$$s'(n) = \eta F[x(n)] \otimes h(n) + z(n) \tag{19}$$

where η is the photodiode responsibility, $F[.]$ is the transformation operation from O-OFDM to PWM, $x(n)$ is the bipolar O-OFDM signal, $h(n)$ is the channel impulse response, $z(n)$ is additive white Gaussian noise with zero mean and variance of δ_z^2, and $\otimes$ denotes the convolution operation. Note that here we have only considered the line of sight (LOS) propagation path in order to illustrate the concept.

As for the recovery of the original OFDM signal at the Rx, the sampling rate of OFDM-PWM must be at least $2R$ times of the ACO-OFDM signal to accurately detect the trailing edge of $s'(n)$. The total number of discrete intervals of $s'(n)$, that is $(N_t + N_c)$, is then estimated from the trailing edge position, and the first half of $x(n)$ is estimated using Eq. (17). Next, the standard ACO-OFDM demodulation process outlined in [3] is adopted to recover the transmitted signal.

4.2. Simulation results

Figure 7 shows the simulated PAPR for ACO-OFDM, ACO-OFDM with classical selected mapping (SLM) technique and the proposed OFDM-PWM method. Here, 16-QAM ACO-OFDM with 256-subcarrier is used as the baseline. The SLM method with six iterations is employed and compared with the proposed OFDM-PWM. **Figure 7** demonstrates that PAPR requirements at CCDF of 10^{-3} are ~3, >12, and < 16 dB for OFDM-PWM, ACO-OFDM with SLM, and ACO-OFDM, respectively. This clearly validates that the proposed OFDM-PWM offers significant PAPR reduction.

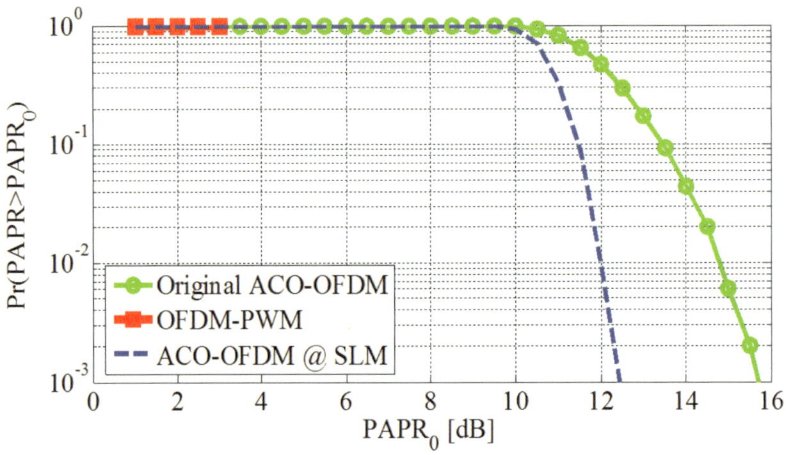

Figure 7. PAPR analysis for 16-QAM ACO-OFDM signals ($N = 256$) with SLM and OFDM-PWM.

Since the average duty cycle of OFDM-PWM is greater than Tc, the proposed OFDM-PWM scheme provides a higher illumination level provided that C is equal to the maximum amplitude of ACO-OFDM sample as shown in **Figure 8**.

Here, we calculate the average optical power of ACO-OFDM and OFDM-PWM to illustrate the illumination level. For the frame shown in **Figure 8**, the average power of OFDM-PWM is ~4 times as much as the original ACO-OFDM frame. Therefore, C can be set to be smaller than the maximum amplitude of ACO-OFDM sample to achieve the desired illumination level and resilience to the $PAPR_0$ limit due to the non-linearity of the LED.

Next, we evaluate the BER performance of the proposed scheme for a line of sight (LOS) VLC link in Matlab environment. The key simulation parameters adopted in this work are summarized in **Table 4**.

Since the pulse-width variation in PWM is linearly proportional to the peak-to-peak amplitude, $v_{\text{p-p-OFDM}}$, of the input signal (i.e., in this case, the O-OFDM), we need to know the time resolution corresponding to the amplitude resolution of the O-OFDM signal. We have selected a nominal value of R as given by:

$$R = \begin{cases} 100, & M \leq 16 \\ \alpha M, & M > 16 \end{cases} \tag{20}$$

where α is an empirical constant, which is set to be four in our simulation, and M is the QAM modulation order. For example, the values of R for 16- and 64-QAM are 100 and 256, respectively.

Figure 9 shows the simulated BER performance against the electrical SNR for OFDM-PWM and ACO-OFDM with 16- and 64-QAM. The simulation results clearly demonstrate that the OFDM-PWM outperforms ACO-OFDM. For example, at a BER of 10^{-4}, the SNR requirements are ~14 and ~18 dB for OFDM-PWM and ACO-OFDM, respectively, for 16-QAM, thus demonstrating an SNR gain of ~4 dB using OFDM-PWM. The SNR gain is higher for the

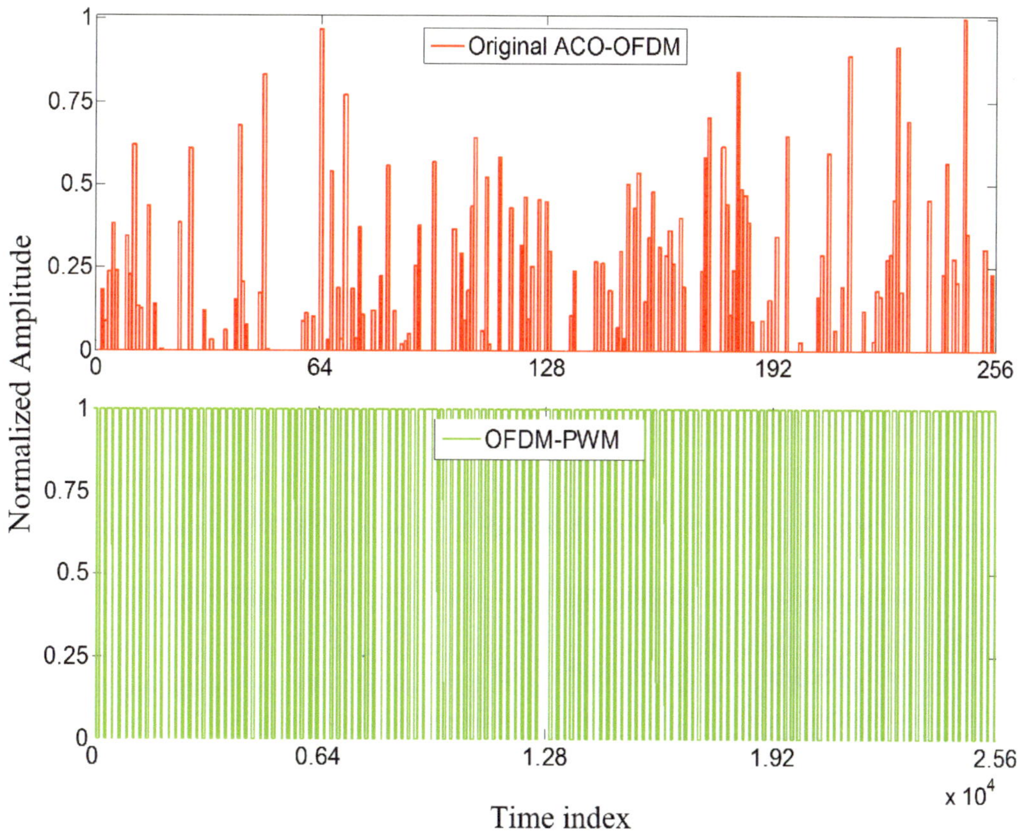

Figure 8. One frame of ACO-OFDM and OFDM-PWM with the same data information ($R = 100$ and 256-subcarriers).

Parameters	OFDM-PWM	ACO-OFDM
Modulation	16/64-QAM	16/64-QAM
No. of subcarriers	256	256
System bandwidth	10 MHz	10 MHZ
Date rate	40/60 Mbit/s	40/60 Mbit/s
Subdivisions	$R = 100/256$	–
Length of frame	25600/65536	256
Iteration	1000	1000

Table 4. Simulation parameters.

higher-order modulation with 64-QAM OFDM-PWM offering a gain of ~9 dB compared to ACO-OFDM.

4.3. Experiment results

The experimental set-up shown in **Figure 10** has been developed to demonstrate the working principle of the proposed concept, validate the simulation results, and compare it with the

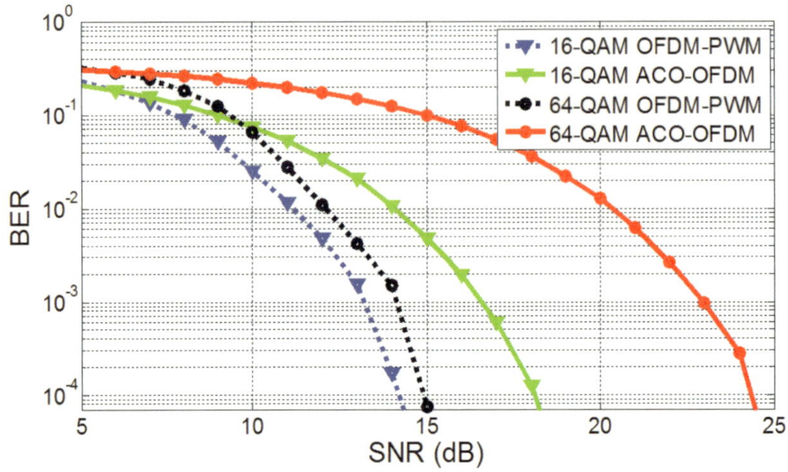

Figure 9. BER performance against the electrical SNR for OFDM-PWM and ACO-OFDM and with 16- and 64-QAM.

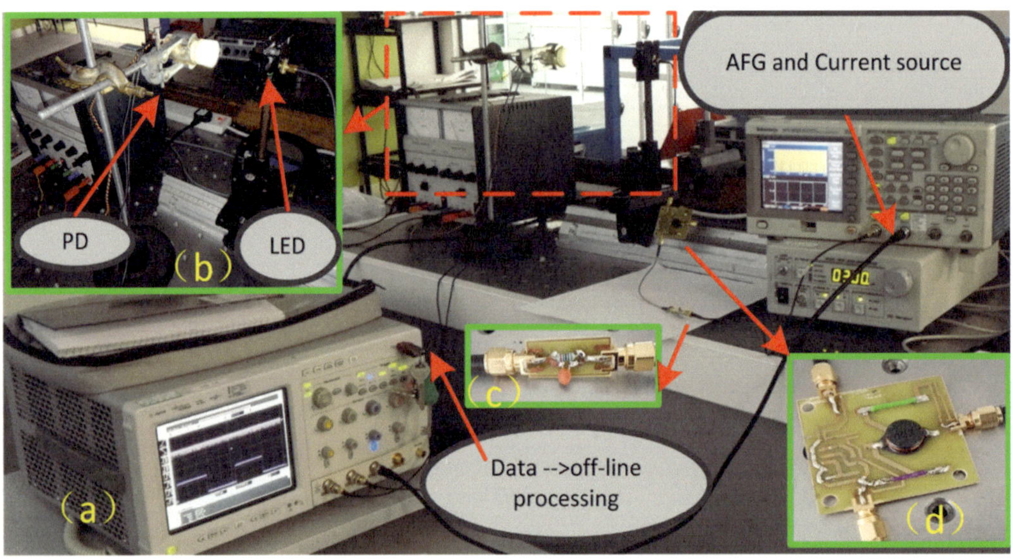

Figure 10. (a). The experimental setup for the proposed scheme. (b) The LED and PD, (c) the RC equalizer circuit, and (d) the DC-bias circuit.

standard ACO-OFDM under the same conditions. The frame frequency and frame sampling rate are shown in **Table 5**.

First, we generate the 16-QAM-based ACO-OFDM and OFDM-PWM signals using the algorithm outlined in Section 4.1. The waveforms are then loaded to the arbitrary function generator (Tektronix AFG 3022) using the LabVIEW 2014 interface. Following this, the output signal is pre-equalized using a simple resistance-capacitance (RC) network, see **Figure 10(c)**, and a DC-bias current I_{dc} of 200 mA is added prior to IM of the LED (Rebel Star 01). Note that

Parameters	OFDM-PWM	ACO-OFDM
Frame frequency f_{frame}	39 kHz	39 kHz
Frame sampling f_{sample}	5 G sample/s	50 M sample/s
DC-bias current	200 mA	200 mA
Modulation	16-QAM	16-QAM
No of random bits	256	256[1]

[1]Only one-fourth of the subcarrier is allocated for information bits in ACO-OFDM.

Table 5. Experimental parameters.

$I_{\text{dc-ACO-OFDM}} < I_{\text{dc-DCO-OFDM}}$, which is mainly used for illumination. Using an RC pre-equalizer network, BLED is increased from ~2 to 12 MHz [29] as shown in **Figure 5(a)**.

In practical VLC systems, LED non-linearity is important and should be considered, which limits the operating dynamic range, $\text{PAPR}_{0,}$ and consequently the SNR. Non-linearity arises due to the imperfection of the driving circuits and the number of emitted photons not being proportional to the injected current in the active region. The measured I-V curve for the LED used in this work is depicted in **Figure 11**, which shows linear and non-linear regions. A lower limit is like the turn-on voltage (TOV) and the saturation voltage.

The receiver's front end consists of an optical concentrator and a photodiode with differential amplifiers, the outputs of which are captured using a digital oscilloscope (Agilent Infinium 40GSa/s) for further offline processing. The oscilloscope sampling rates are set to 50 M and 5 G sample/s for ACO-OFDM and OFDM-PWM, respectively. Signals synchronization, down-sampling, and signal recovery are carried out offline. We then evaluate the BER performance of both ACO-OFDM and OFDM-PWM systems under the same conditions. For higher-order modulations (i.e., 64-QAM), the generated digital OFDM-PWM's length is 65, 536, which is beyond the 32 K memory of AFG 3022. Therefore, we could only transmit one frame of 16-QAM-based 256-subcarrier OFDM-PWM due to the limited memory of AFG3022. The received waveforms of ACO-OFDM and OFDM-PWM are shown in **Figure 12(a)** and **(b)**, respectively. The square waveforms shown are used to indicate the period of the OFDM signals. Following the offline processing, the original ACO-OFDM signal is fully recovered as shown in **Figure 12(c)**.

Figure 13 shows the estimated BER performance, which is obtained experimentally against the peak-to-peak amplitude, Vp-p values of 16-QAM-based ACO-OFDM and OFDM-PWM signals. Due to a limited number of transmitted bits, it is not feasible to accurately measure the BER below the FEC limit of 10^{-3}. Nonetheless, the graph clearly demonstrates the advantage of OFDM-PWM over ACO-OFDM, where it displays an improved BER performance in both linear and non-linear regions of the LED. For Vp-p < 2 V, the BER floor for OFDM-PWM is $<5 \times 10^{-3}$, whereas for ACO-OFDM, it is $> 10^{-1}$, except at Vp-p of 2, where the BER is 10^{-2}. For Vp-p > 2 V, the BER performance shows degradation for both schemes. However, OFDM-PWM is still superior to ACO-OFDM. Since the information is carried in the pulse width, rather than pulse amplitude as in ACO-OFDM, OFDM-PWM displays a higher resiliency to the non-linearity of the LED and clipping, thus offering an improved performance.

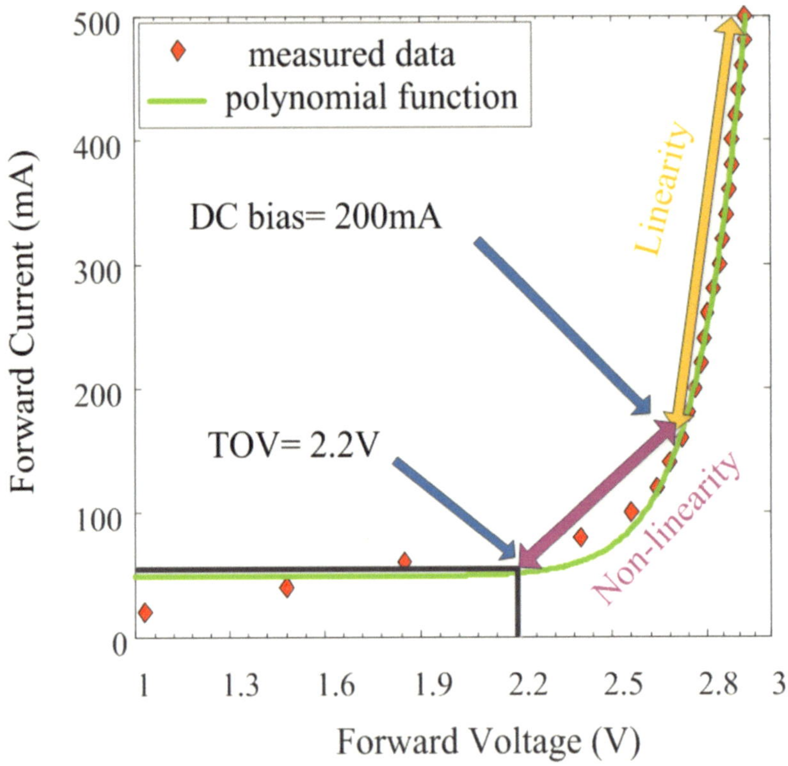

Figure 11. The LED I-V plot using polynomial function (Rebel Star 01, LED). TOV: turn-on voltage.

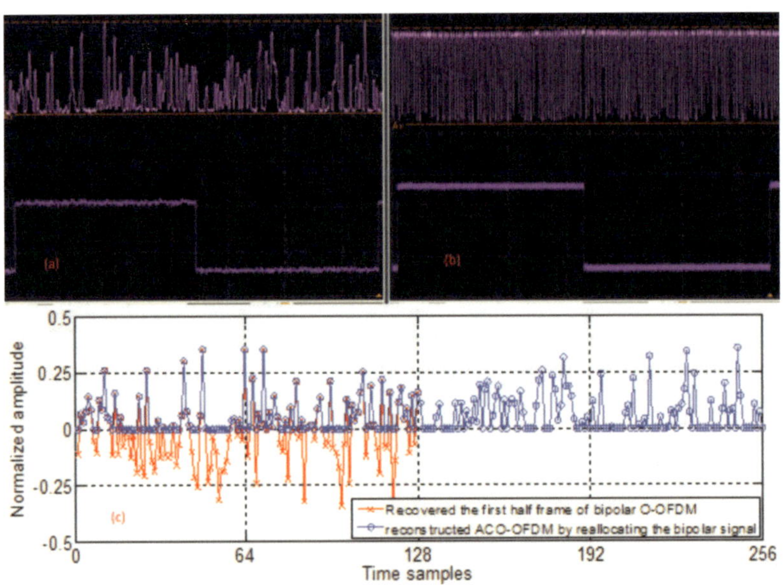

Figure 12. Waveforms of a single frame of 256-subcarrier for: (a) ACO-OFDM, (b) OFDM-PWM with $R = 100$, and (c) the recovered half frame of bipolar signal and the reconstructed ACO-OFDM frame. Note the square wave in (a) and (b) is to show the period of OFDM waveforms.

http://dx.doi.org/10.5772/intechopen.68763

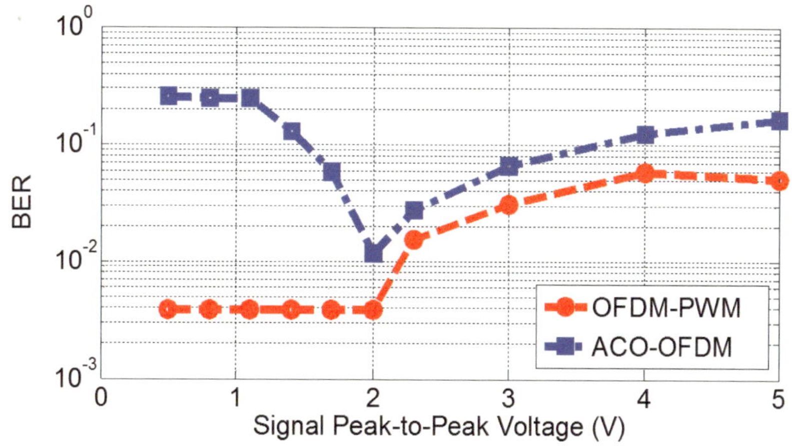

Figure 13. Experimental BER against the peak-to-peak amplitude for 16-QAM OFDM-PWM and ACO-OFDM under the same conditions.

5. Conclusion

In this chapter, we proposed two novel PAPR reduction schemes for O-OFDM-based VLC systems. The first scheme is implemented in the time domain to mitigate the high PAPR experienced in ACO-OFDM, which uses the Gaussian blur method with a Toeplitz matrix-based Gaussian kernel function at the transmitter. We have shown that for 256-subcarrier ACO-OFDM, a ~6dB improvement in PAPR is achieved in terms of CCDF compared with the original ACO-OFDM. By taking advantage of the special time-domain structure of ACO-OFDM symbols, the OMP algorithm is employed to efficiently recover the original ACO-OFDM signal at the receiver. Simulation results have shown that a ~6 dB PAPR reduction requires only a ~2 dB of additional SNR compared to the ideal ACO-OFDM scheme at a BER of 10–4 and with the signal reconstruct error less than 0.1. Then, the second scheme is an improved modulation scheme based on the conversion of an ACO-OFDM signal to PWM. The OFDM-PWM offers significant advantages compared with the traditional ACO-OFDM scheme including lower PAPR, higher luminance, improved BER performance, and enhanced resiliency to the source non-linearity. The simulation and experimental results both show that a significant BER improvement can be achieved with OFDM-PWM compared to ACO-OFDM.

Author details

Tian Zhang[1], Jun Yao[2] and Shuxu Guo[1]*

*Address all correspondence to: guosx@jlu.edu.cn

1 College of Electronic Science and Engineering, Jilin University, Changchun, China

2 Department of Electrical and Computer Engineering, Carnegie Mellon University, Pittsburgh, PA, USA

References

[1] Ghassemlooy Z, Arnon S, Uysal M, Xu Z, Cheng J. Emerging optical wireless communications-advances and challenges. IEEE Journal on Selected Areas in Communications. 2015;**33**(9):17381749/15381761. DOI: 10.1109/JSAC.2015.2458511

[2] Minh HL, O'Brien D, Faulkner G, Zeng L, Lee K, Jung D, Oh Y, Won ET. 100-Mb/s NRZ visible light communications using a postequalized white LED. IEEE Photonics Technology Letters. 2009;**21**(15):10631065/10779028. DOI: 10.1109/LPT.2009.2022413

[3] Dissanayake SD, Armstrong J. Comparison of ACO-OFDM, DCO-OFDM and ADO-OFDM in IM/DD systems. Journal of Lightwave Technology. 2013;**31**(7):10631072/13287715. DOI: 10.1109/JLT.2013.2241731

[4] Kimura H, Asaka K, Nakamura K, Kimura S, Yoshimoto N. Energy efficient IM-DD OFDM-PON using dynamic SNR management and adaptive modulation. Optics Express. 2014;**22**(2):17891795. DOI: https://doi.org/10.1364/OE.22.001789

[5] Jiang T, Wu Y. An overview: Peak-to-average power ratio reduction techniques for OFDM Signals. IEEE Transactions on Broadcasting. 2008;**54**(2):257268/10002861. DOI: 10.1109/TBC.2008.915770

[6] Beril Inan SC, Lee J, Randel S, Neokosmidis I, Koonen AMJ, Walewski JW. Impact of LED nonlinearity on discrete multitone modulation. Journal of Optical Communications and Networking. 2009;**1**(5):439451. DOI: https://doi.org/10.1364/JOCN.1.000439

[7] Mesleh R, Elgala H, Haas H. LED nonlinearity mitigation techniques in optical wireless OFDM communication systems. Journal of Optical Communications and Networking. 2012;**4**(11):865875. DOI: https://doi.org/10.1364/JOCN.4.000865

[8] O'Neill R, Lopes LB. Envelope variations and spectral splatter in clipped multicarrier signals. Toronto: IEEE; 1995; p. 7175. DOI: 10.1109/PIMRC.1995.476406

[9] Kang W, Hranilovic S. Power reduction techniques for multiple-subcarrier modulated diffuse wireless optical channels. IEEE Transactions on Communications. 2008;**56**(2):279288/9836375. DOI: 10.1109/TCOMM.2008.060609

[10] You R, Kahn JM. Average power reduction technique for multiple subcarrier intensity modulated optical signals. IEEE. 2000; pp. 1620–1627, New Orleans, LA, USA. DOI: 10.1109/ICC.2000.853769

[11] Nadal L, Moreolo MS, Fabrega JM, Junyent G. Low complexity bit rate variable transponders based on optical OFDM with PAPR reduction capabilities. Spain: IEEE; 2012; p. 16. DOI: 10.1109/NOC.2012.6249918

[12] Popoola WO, Ghassemlooy Z, Stewart BG. Pilot-assisted PAPR reduction technique for optical OFDM communication systems. Journal of Lightwave Technology. 2014;**32**(7):13741382/14111646. DOI: 10.1109/JLT.2014.2304493

[13] Wasiu O. Popoola; Zabih Ghassemlooy; Brian G. Stewart. Optimizing OFDM based visible light communication for high throughput and reduced PAPR. In: 8–12 June 2015; IEEE; 2015. p. 1322–1326, London, UK. DOI: 10.1109/ICCW.2015.7247361

[14] Zhang T, Ghassemlooy Z, Ma C, Guo S. PAPR reduction scheme for ACO-OFDM based visible light communication systems. Optics Communications. 2017;**383**:7580. DOI: http://dx.doi.org/10.1016/j.optcom.2016.07.073

[15] Flusser J, Farokhi S, Höschl C, Suk T, Zitová B, Pedone M. Recognition of images degraded by Gaussian Blur. IEEE Transactions on Image Processing. 2016;**25**(2):790806/15746081. DOI: 10.1109/TIP.2015.2512108

[16] Zhang T, Ghassemlooy Z, Rajbhandari S, Popoola WO, Guo S. OFDM-PWM scheme for visible light communications. Optics Communications. 2017;**385**:213–218. DOI: http://dx.doi.org/10.1016/j.optcom.2016.10.038

[17] Guerra V, Suarez-Rodriguez C, El-Asmar O, Rabadan J, Perez-Jimenez R. Pulse width modulated optical OFDM. IEEE. 2015; pp. 1333–1337, London, UK. DOI: 10.1109/ICCW.2015.7247363

[18] Armstrong J. OFDM for optical communications. Journal of Lightwave Technology. 2009;**27**(3):189204/10479055. DOI: 10.1109/JLT.2008.2010061

[19] Li F, Wu K, Zou W, Chen J. Optimization of LED's SAHPs to simultaneously enhance SNR uniformity and support dimming control for visible light communication. Optics Communications. 2015;**341**:218–227. DOI: http://dx.doi.org/10.1016/j.optcom.2014.12.025

[20] Rebollo-Neira L, Lowe D. Optimized orthogonal matching pursuit approach. IEEE Signal Processing Letters. 2002;**9**(4):137140/7276019. DOI: 10.1109/LSP.2002.1001652

[21] Han SH, Lee JH. An overview of peak-to-average power ratio reduction techniques for multicarrier transmission. IEEE Wireless Communications. 2005;**12**(2):5665/8389285. DOI: 10.1109/MWC.2005.1421929

[22] Emmanuel J Candes, Michael B Wakin. An introduction to compressive sampling. IEEE Signal Processing Magazine. 2008;**25**(2):2130/9903092. DOI: 10.1109/MSP.2007.914731

[23] Davenport MA, Wakin MB. Analysis of orthogonal matching pursuit using the restricted isometry property. IEEE Transactions on Information Theory. 2010;**56**(9):43954401/11473346. DOI: 10.1109/TIT.2010.2054653

[24] Tony Cai T, Wang L. Orthogonal matching pursuit for sparse signal recovery with noise. IEEE Transactions on Information Theory. 2011;**57**(7):46804688/12068809. DOI: 10.1109/TIT.2011.2146090

[25] Xu W, Wu M, Zhang H, You X, Zhao C. ACO-OFDM-specified recoverable upper clipping with efficient detection for optical wireless communications. IEEE Photonics Journal. 2014;**6**(5):117/14580758. DOI: 10.1109/JPHOT.2014.2352643

[26] Khalid AM, Cossu G, Corsini R, Choudhury P, Ciaramella E. 1-Gb/s transmission over a phosphorescent white LED by using rate-adaptive discrete multitone modulation. IEEE Photonics Journal. 2012;**4**(5):14651473. DOI: 10.1109/JPHOT.2012.2210397

[27] Armstrong J, Schmidt BJC. Comparison of asymmetrically clipped optical OFDM and DC-biased optical OFDM in AWGN. IEEE Communications Letters. 2008;**12**(5):343345/9966259. DOI: 10.1109/LCOMM.2008.080193

[28] Wilson W, Ghassemlooy Z. Pulse time modulation techniques for optical communications: a review. IEE Proceedings J-Optoelectronics. 1993;**140**(6):347357/4587345. DOI: 10.1049/ip-j.1993.0055

[29] Haigh PA, Ghassemlooy Z, Minh HL, Rajbhandari S, Arca F, Tedde SF, Hayden O, Papakonstantinou I. Exploiting equalization techniques for improving data rates in organic optoelectronic devices for visible light communications. Journal of Lightwave Technology. 2012;**30**(19):30813088/13014650. DOI: 10.1109/JLT.2012.2210028

Chapter 6

Collaborative VLC/IROW Systems

Ahmed Taha Hussein,
Mohammed T. Alresheedi and
Jaafar M. H. Elmirghani

Additional information is available at the end of the chapter

http://dx.doi.org/10.5772/intechopen.68474

Abstract

Dimming is an important feature of an indoor lighting system where the illumination level can be controlled by the user. Therefore, integrating a visible light communication (VLC) system with an illumination system poses some challenges. One of the main issues is that the light unit should be "ON" all the time to ensure continuous communication. To ensure acceptance and adoption of VLC systems, an important issue should be addressed: how to communicate when the lights are "OFF" or partially dimmed. In this chapter, we propose five new infrared optical wireless (IROW) systems to support VLC systems when the light is totally turned off or significantly dimmed. To take advantage of both VLC and IROW, we introduce and implement the concept of a collaborative VLC/IROW system. In addition, we investigate the impact of partial dimming on the VLC system's performance, and we propose an adaptive rate technique (ART) to mitigate the impact of light dimming. Moreover, in the case of no dimming, the VLC and IROW systems can collaborate to increase the data rate so it is higher than that in the pure VLC system. We have achieved 10 Gbps in an indoor environment, which is a 2× increase in the data rate compared with a pure VLC system.

Keywords: collaborative VLC/IROW, delay spread, SNR

1. Introduction

A visible light communication (VLC) system has the potential to become a complementary technology to its radio frequency (RF) counterpart, and this is due to the hundreds of THz of license free bandwidth, high security, energy efficiency and immunity to electromagnetic interference [1]. Despite the advantages presented by the VLC medium, there are several challenges facing VLC systems to achieve high data rates. These challenges include the low modulation bandwidth of the light-emitting diodes (LEDs) and inter-symbol interference (ISI) due to diffuse transmission where the optical signal reaches the receiver through a number of different paths that result in pulse spread, which in turn leads to ISI. Various techniques have been proposed to mitigate the limitations (low modulation bandwidth of LEDs and ISI) in VLC systems to achieve high data rates. On-going research activities intend to increase the data rates of indoor VLC systems by replacing LEDs with visible laser diodes (LDs) coupled with the use of an imaging receiver instead of the conventional wide field of view (FOV) receiver [2–5].

The concept of VLC systems is based on the use of light units (LEDs/LDs) for both lighting and communications. Therefore, using light units (i.e. LED or LD) for communications should not interfere with the light units' main function (i.e. illumination). A user may arbitrarily dim the light source in the VLC system to save power, so it is essential to maintain communication in this case. An infrared optical wireless (IROW) system can be used to collaborate with the VLC system under partial/full dimming. IROW systems can provide high transmission rates similar to VLC systems [6, 7]. This is due to the wider modulation bandwidth of the laser sources used in IROW systems instead of white LEDs. IROW systems can use simple signal processing functionality and simple modulation formats while having a much higher bandwidth available for future usage. IROW systems have some additional advantages compared to VLC. For example, light dimming is not an issue in IR systems, and the uplink implementation using IR is convenient as it avoids bright visible light next to the user's equipment, next to a laptop, for example. However, IROW systems have two major challenges: multipath dispersion due to reflections and sensitivity to additive shot noise coming from artificial background lighting or sunlight [6, 8].

The received power at the VLC receiver is reduced when the user dims the light to low levels, and this will lead to a degradation in the signal-to-noise ratio (SNR) and affect the data rate achievable. Recently, hybrid schemes were proposed to support VLC systems. RF-based systems are used to supplement the VLC system [9, 10]. However, achieving a high transmission rate (multi gigabits per second) and security is the most challenging parts. Therefore, the VLC link needs to collaborate with the IROW connection to provide continuous data transmission. When the VLC link has recovered (i.e. there is no dimming), the VLC and IROW systems can cooperate together to increase the data rate at the receiver side.

In this chapter, we addressed the weakness of VLC systems and provide practical solutions when the light is partially dimmed or totally switched off. Subsequently, we have achieved data rates more than those reported in Refs. [9, 10]. In this study, we report the use of infrared systems that utilise a LD source to support the VLC system when the light is totally switched off. IR optical communication has the same advantages as VLC systems. It can also provide high transmission data rates similar to VLC systems and potentially higher data rates (data rates up to 15 Gbps can be achieved) [11]. In addition, we investigate the performance of a VLC system under the impact of different levels of dimming, propose an adaptive rate technique (ART) and produce the concept of cooperation between VLC and IROW systems. The VLC system is able to achieve high data rates (5 Gbps) when the light units are "ON." However, the achieved data rate (i.e. 5 Gbps) will decrease as a result of light dimming. Therefore, IROW systems are proposed to ensure the continuity of the wireless communication and to maintain the target data rate (5 Gbps), as the IROW system can be used to compensate for the degradation of the data rate due to dimming in the VLC system. In addition, the IROW system can be used to increase the data rates so they are higher than the target (5 Gbps) when the lights are "ON," and the VLC system is operating normally.

Five IROW systems are proposed: hybrid diffuse IR (HDIR) with wide field of view receiver, HDIR with imaging receiver, beam steering IR (BSIR) with imaging receiver, cluster distributed IR (CDIR) with imaging receiver and a cluster distributed beam steering IR (CDBSIR) system, to collaborate with the VLC system. The data rates achieved by our proposed backup systems are 2.5 and 5 Gbps when using a very simple modulation format (on-off keying, OOK) and avoid the use of relatively complex wavelength division multiplexing approaches. In this study, we used two types of receivers: wide FOV receiver and an imaging receiver with selective combining (SC) to choose the best pixel.

The remainder of this chapter is organised into the following sections: Section 2 presents the simulation environment and VLC/IROW channel model. Section 3 presents the proposed systems' configurations. Section 4 introduces the ART and the impact of dimming on the VLC system performance. Section 5 introduces the simulation results and discussion of the IROW systems. Section 6 provides the simulation results and discussion of the collaborative VLC/IROW system in an empty room. Finally, conclusions are drawn in Section 7.

2. Simulation environment and VLC/IROW channel model

To study the benefits of our proposed systems in an indoor environment, a simulation based on a ray tracing algorithm was performed in an unoccupied room (empty) with 8 m × 4 m (length × width) floor dimensions and a height of 3 m to the ceiling. The walls and ceiling were segmented into small reflective elements (dA). The reflective elements were treated as small secondary emitters that diffuse the received signal in the shape of a Lambertian pattern with $n = 1$, where n is the Lambertian emission order, having a reflectivity of 0.8 for the walls and ceiling and 0.3 for the floor [12–15]. Element sizes of 5 cm × 5 cm for the first order reflections and 20 cm × 20 cm for the second order reflections are employed for all arrangements. Previous work has studied the received optical power within an indoor environment. They found that most of the received optical power is located within the two first-order reflections (1st and 2nd). Third order and higher reflections are highly attenuated [16]. Hence, two bounces are considered in our calculations.

In indoor OW communication systems, intensity modulation with direct detection (IM/DD) is the preferred choice as a result of its reduced cost and complexity [17]. The receiver makes use of a detector that produces a photocurrent $I(t)$ that is proportional to the received instantaneous optical power. The indoor optical wireless IM/DD channel can be given by [18]:

$$I(t, Az, El) = \sum_{m=1}^{M_t} Rx(t) \otimes h_m(t, Az, El) + \sum_{m=1}^{M_t} Rn_m(t, Az, El) \tag{1}$$

where t is the absolute time and El and Az represent the direction of the arrival in the elevation and azimuth angles, respectively. M_t is the total number of reflecting elements, $x(t)$ is the transmitted instantaneous optical power, ⊗ denotes convolution, R is the detector responsivity and $I(t, Az, El)$ is the received photocurrent at a certain location resulting from M_t reflecting surfaces. Lastly, $n_m(t, Az, El)$ represents the background light noise due to m^{th} reflecting elements at the receiver. Many parameters can be obtained from the simulated impulse response, for example, 3 dB channel bandwidth, the delay spread and SNR. The delay spread of an impulse response is given by [6]:

$$D = \sqrt{\frac{\sum (t_i - \mu)^2 P_{ri}^2}{\sum P_{ri}^2}} \tag{2}$$

where t_i is the delay time associated with the received optical power P_{ri}, and μ is the mean delay given by:

$$\mu = \frac{\sum t_i P_{ri}^2}{\sum P_{ri}^2} \tag{3}$$

1. The 3 dB channel bandwidth is equal to the frequency when the magnitude response falls by 3 dB.

2. The SNR of the received signal can be calculated by taking into account the powers associated with logic 0 and logic 1 (P_{S0} and P_{S1}), respectively. The SNR is given by [18]:

$$SNR = \left(\frac{R(P_{s1} - P_{s0})}{\sigma t}\right)^2 \tag{4}$$

where R is the receiver responsivity and σ_t is the standard deviation of the total noise that is the sum of the shot noise, thermal noise and signal dependent noise. σ_t can be calculated as follows:

$$\sigma t = \sqrt{\sigma_{shot}^2 + \sigma_{preamplifier}^2 + \sigma_{signal}^2} \tag{5}$$

where σ_{shot}^2 represents the background shot noise component, $\sigma_{preamplifier}^2$ represents the preamplifier noise component, and σ_{signal}^2 represents the shot noise associated with the received signal [19]. Bit rates of 2.5 and 5 Gbps are evaluated in our proposed systems. We used the preamplifier design proposed in Ref. [20]. In this study, we considered the SC method of processing the electrical signal from different pixels in an imaging receiver. In the SC, the receiver simply selects the pixel with the largest SNR among all the pixels. The SNR_{SC} is given by Ref. [18]:

$$SNR_{SC} = Max_i \left(\frac{R(P_{s1} - P_{s0})_i}{\sigma t_i}\right)^2 \quad 1 \leq i \leq j \tag{6}$$

where j represents the number of pixels ($j = 50$ in our imaging receiver).

3. Proposed systems' configurations

In this section, proposed OW systems are presented, analysed and compared to identify a reliable and high data rate wireless communication system for an indoor user.

3.1. Imaging LD-VLC System

The imaging LD-VLC system used eight lighting fixtures (i.e. RGB-LDs) on the ceiling connected by fibre optic cable and controlled by a central controller. The imaging LD-VLC system was introduced in Ref. [2], and it is considered here to investigate its performance under the impact of different levels of dimming. In addition, it will be integrated with an IROW system to provide reliable and high data rate services for an indoor user. **Figure 1** shows the architecture of the imaging LD-VLC system.

3.2. IROW Systems

In this section, five IROW systems are presented, analysed and compared to identify the most appropriate system for use to collaborate with the VLC system (imaging LD-VLC).

3.2.1. Hybrid diffuse IR (HDIR) system employing wide FOV receiver

The HDIR employed one IR transmitter located at the centre of the ceiling, which can provide a direct LOS link at the receiver on the communication plane (CP). **Figure 2** shows the HDIR communication architecture. In this case, most of the power is gathered from the LOS link, and the rest of power is collected through reflections. The proposed system (HDIR) uses a

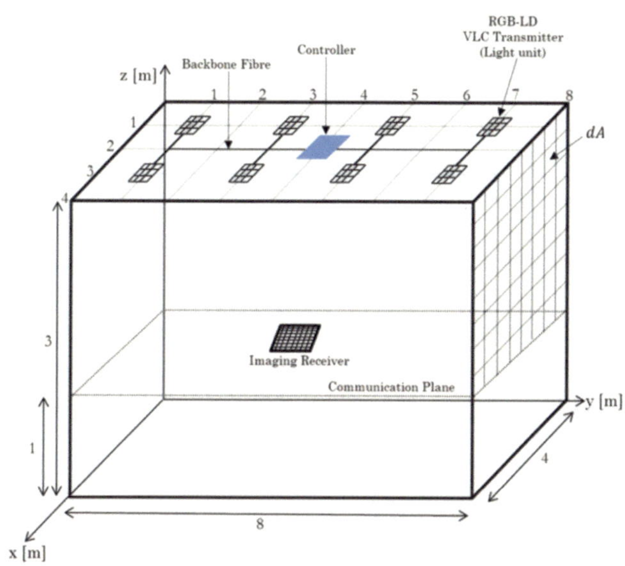

Figure 1. Architecture of imaging LD-VLC system with imaging receiver on communication plane.

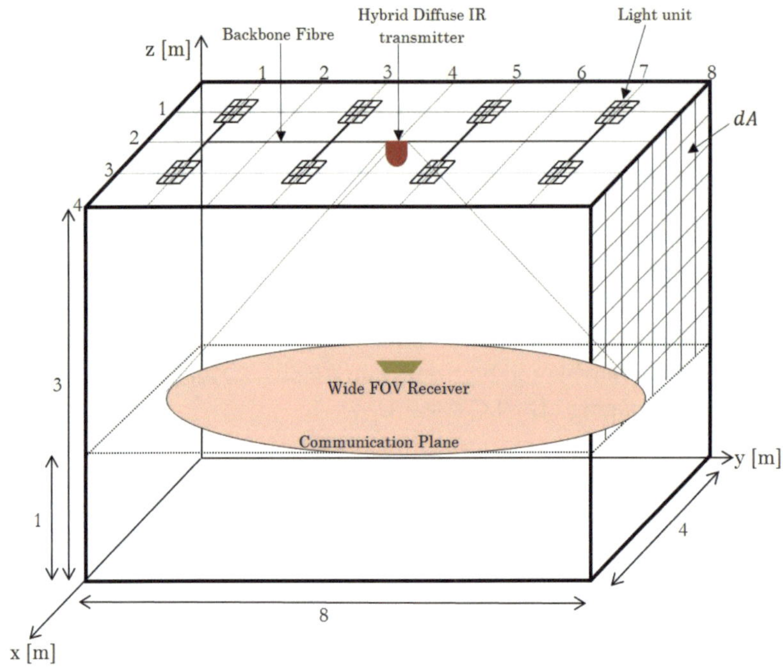

Figure 2. Architecture of HDIR with wide FOV receiver on communication plane.

single-wide beam source, typically with a Lambertian pattern where the transmitted optical signal fully diffuses over the environment. The IR transmitter is connected to all the visible light sources via fibre links (to link to main network in the building) and simple control circuits (located at the centre of the room). When the light is dimmed or the received optical power falls below a certain threshold, a feedback IR signal at a low rate is sent by the receiver to the controller to switch the link into the backup system (i.e. HDIR). The receiver consists of the VLC and IR detectors connected through an electronic switch to control their functions. In this system, we used a conventional single-element wide FOV (90°) photodetector with photo sensitive area of 1 mm^2. The HDIR transmitter is positioned at the centre of the room at (2, 4, and 3 m), is pointed downwards, and emits 1 W with an ideal diffuse pattern. Exposure to optical radiation at such power levels can be hazardous to the skin and eyes. Nevertheless, different techniques can be used to reduce the impact of the high laser power, such as extending the source size, destroying its spatial coherence using holograms mounted on the transmitter or the use of arrays of transmitters. Pohl et al. have shown that such a source may use an integrating sphere as a diffuser to emit optical power in the range from 100 mW to 1 W [21]. Therefore, a transmitter power of 1 W will be assumed in this system.

3.2.2. Hybrid diffuse IR (HDIR) system employing imaging receiver

The HDIR system with imaging receiver has a comparable room configuration and uses the same IR transmitter as the former system. However, a 50 pixels imaging receiver is used here. The imaging receiver shows two main pros over the traditional non imaging receivers. Firstly, all detectors share a common concentrator (e.g. a lens). Hence, it can be fabricated with a smaller size and lower cost. Secondly, all photodetectors can be placed on a single plane. Therefore, the designer can opt to use a larger number of detectors with small detector areas and narrow FOV. This will improve the receiver bandwidth (the input capacitance can be reduced by reduce photodetector area) and mitigate the effect of multipath dispersion. In this chapter, we employed the imaging receiver design proposed in Ref. [2].

3.2.3. Beam steering IR (BSIR) system employing imaging receiver

In contrast to the HDIR, in the BSIR system, the IR transmitter uses beam steering to direct the IR beam toward the receiver position. The IR transmitter faces downward at the centre of the ceiling as shown in **Figure 3**.

It has been proved that beam steering is an effective approach that can help to maximise the receiver's SNR, regardless of the receiver's FOV, the receiver's orientation and the transmitter's position [5]. However, this technique requires intensive time and calculations on the processor to generate a hologram at each step. In our new BSIR system, we propose an efficient stored vocabulary hologram method for beam steering in our backup system. For a large room of 4 m × 8 m, the floor (i.e. CP) is divided into 512 regions (0.25 m × 0.25 m per region). The total number of holograms to be stored in our design is *N*, where *N* represents the number of regions into which the CP is divided. This large number of regions has been chosen to accurately identify the receiver location during its motion (user mobility). Since the

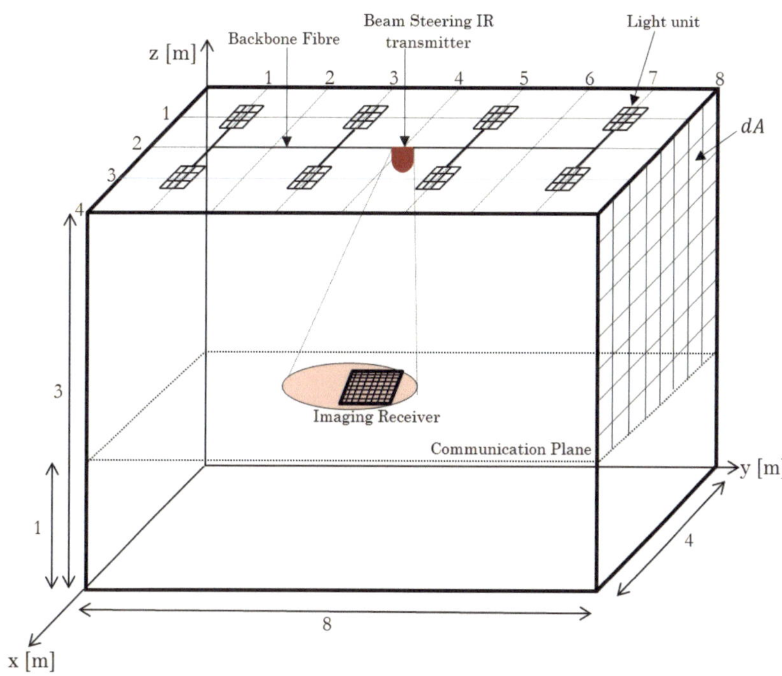

Figure 3. Architecture of BSIR with imaging receiver on communication plane.

transmitter is fixed at the centre of the room (close to the controller), the transmitter uses a hologram that steers the beam (narrow direct LOS link) to the optimum location if the receiver exists in any one of the regions. The idea of finite vocabulary stored holograms has been recently proposed in VLC systems [5], and it is developed here for the first time in the BSIR system to improve the performance of the HDIR system.

In the BSIR system, the transmitter emitted 100 mW, where the hologram used $n = 45$ to focus the beams within a 0.25 m × 0.25 m area. A small transmitted power (100 mW) is used in our proposed system to satisfy the eye safety regulations in the indoor OW environment.

3.2.4. Cluster distributed IR (CDIR) system employing imaging receiver

In the CDIR system, IR transmitters are used that utilise LD sources to support the VLC systems when the light is partially dimmed or totally switched off. The CDIR system employs more than one IR source and distributes them on the ceiling, that is, each IR source is attached to a VLC transmitter (i.e. light unit). All IR sources are connected via fibre and a control unit to perform cluster mechanisms. The new concept of using IR clusters is employed to design a new geometry that can achieve a good performance in mobile IR communications.

A custom design for the imaging receiver (similar to the one in the previous systems) is used to reduce the impact of multipath dispersion and ISI. **Figure 4** shows the architecture of our CDIR system. The proposed system consists of eight IR sources, and each is attached to a visible light source located on the ceiling, which can provide a direct LOS link to the receiver on the CP. The IR transmitters are connected to all visible light sources through fibre links (to link to the main network in the building) and simple control circuits. As shown in **Figure 4**,

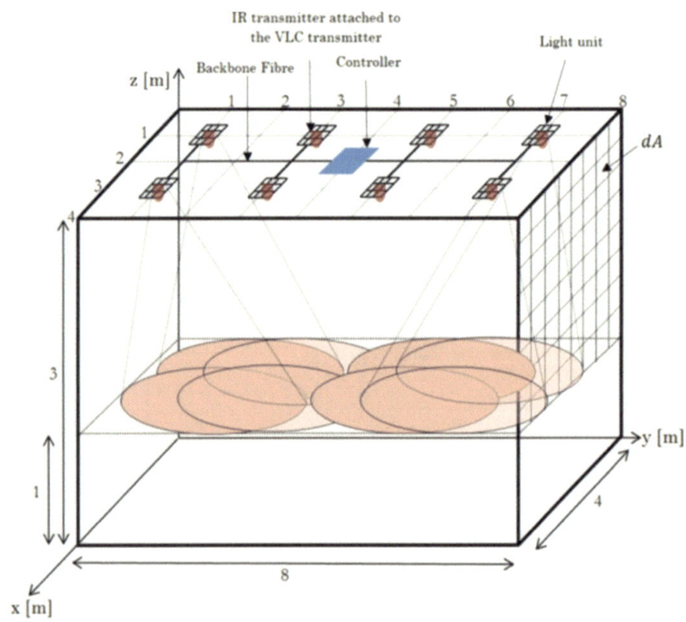

Figure 4. Architecture of CDIR when all IR transmitters are "ON".

each IR source forms part of a cluster that can cover over 2 m × 2 m. When the light is dimmed or the received optical power falls below a certain threshold, the receiver sends a feedback signal at a low rate to the VLC transmitter to switch the link into the supporting CDIR system. A select the best (STB) algorithm is used (similar to the ones used in Ref. [22]) to select the optimum link between the IR transmitter and receiver (under mobility, this algorithm can be called periodically). The first step is to switch ON each IR source individually. Each source uses a single-wide beam, typically with a Lambertian pattern with $n = 1$, where n is the Lambertian emission order (the transmitted optical signal fully diffuses over the environment). The receiver then computes the received power and the SNR associated with each source. The receiver sends a signal, at a low rate, back to the control unit conveying the information about the SNR weight associated with each source to identify the receiver location (the receiver is located near to the source that has the highest SNR). The transmitter switches ON only the optimal IR source (source that is nearest to the receiver location) and other sources remain off to reduce the impact of multipath dispersion.

3.2.5. Cluster distributed beam steering IR (CDBSIR) system

In contrast to the CDIR, in the CDBSIR system, the IR transmitter uses the beam steering technique to steer the IR beam towards the receiver location. Like CDIR, the CDBSIR system employs a STB algorithm to select the closest IR transmitter to the receiver. The selected IR transmitter in the STB algorithm will then apply a finite stored hologram in the system, instead of a diffuse source (as in the CDIR system). The stored hologram in the proposed system produces an IR beam to scan an area of 2 m × 2 m and steer the IR beam towards the receiver location. The floor (2 m × 2 m) under the IR source is subdivided into small areas; for example, we divided it to 256 subdivisions (0.125 m × 0.125 m) as shown in **Figure 5**. The IR transmitter divides the stored holograms into four quadrants with a boundary based on the hologram transmission angles ($-\delta_{min}$ to 0) and (0 to δ_{max}) in both x and y axes. The transmitter first tests the middle hologram at each quadrant (four holograms will be initially tested) to identify the sub-optimal quadrant; hence, this will reduce the number of holograms that need to be tested by a factor of four in the first step. The receiver sends a feedback signal at a low rate that informs the transmitter about the SNR associated with each hologram. The hologram that results in the best receiver SNR is identified as a sub-optimum hologram, and the quadrant that includes this sub-optimum hologram will be divided in the next step into four sub-quadrants. A number of iterations are carried out until the final optimum location is identified. The concept of beam steering has been recently proposed in a VLC system [5], and it is developed here to improve the performance of the CDIR system.

3.3. Collaborative VLC/IROW System

It is desirable to continue to provide a high data rate service, while a user dims the light source to any level. However, the received power at the VLC receiver is reduced when the user dims the light to low levels, and this leads to a degradation in the SNR and affects the achievable data rate. Therefore, a collaborative VLC/IROW system is introduced to address this issue, and when the VLC has partial dimming, such as 75 or 50%, both the IROW and VLC systems can collaborate to maintain the target data rate (5 Gbps). The IROW system can be used to

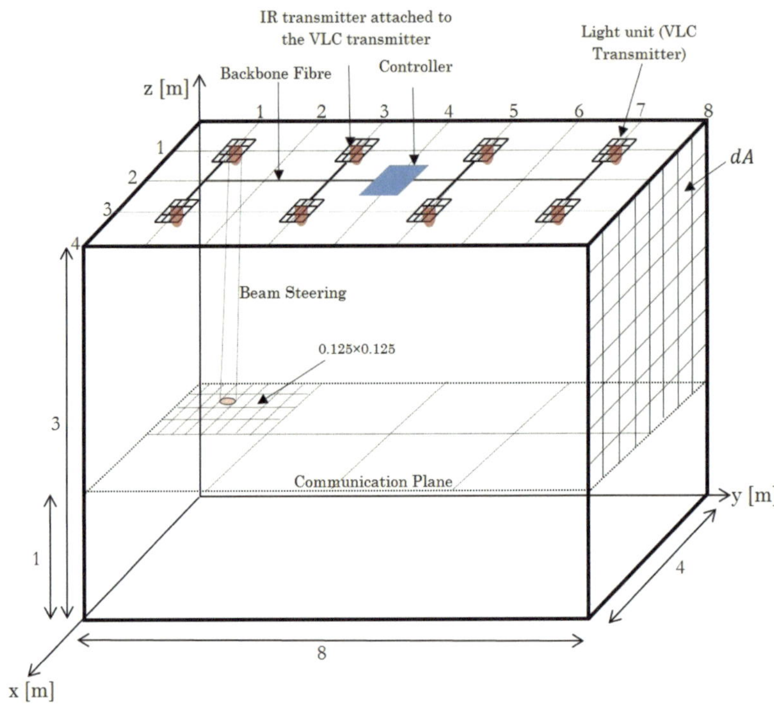

Figure 5. Architecture of CDBSIR system.

compensate for the degradation of the data rate due to dimming in the VLC system. It should be noted that the IROW system sends information at a fixed rate of 5 Gbps. In the case of no dimming, the VLC and IROW can be used to increase the data rate higher than the maximum VLC data rate (i.e. higher than 5 Gbps); hence, the achieved data rate will be 10 Gbps instead of 5 Gbps. **Figure 6** shows the architecture of the VLC/IROW system. The proposed system consists of eight IR sources (similar to those used in the CDBSIR system) and eight VLC transmitters (similar to those used in the imaging LD-VLC system) and employs an imaging receiver with 50 pixels. In order to eliminate interference between uplink and downlink channels, the downlink IR channel is used 850 nm and the IR uplink is used 1550 nm.

4. Adaptive rate technique

In this section, we introduce the ART and evaluate the performance of the imaging LD-VLC system under the impact of multiple levels of light dimming (25, 50 and 75%). The results are presented in terms of the SNR at different operating data rates (5, 2.5 and 1.25 Gbps).

It should be noted that an SNR equal to 13.54 dB is needed for a 10^{-6} probability of error (BER). Therefore, we have chosen BER = 10^{-6} as the threshold in the imaging LD-VLC, and we employ an ART to ensure that we have an acceptable quality communication link under different levels of dimming.

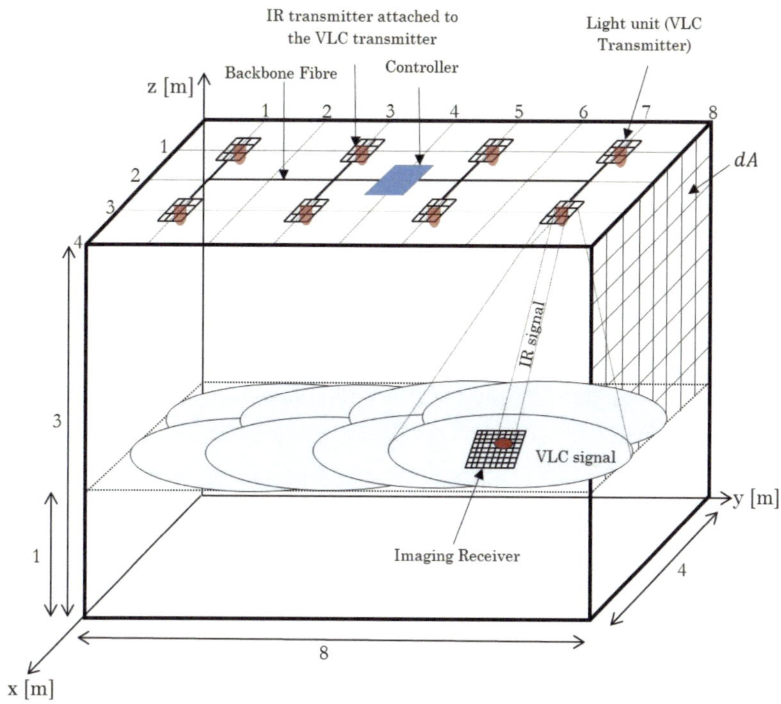

Figure 6. Architecture of VLC/IROW collaborative system.

ART is carried out at the receiver and the controller. First, the receiver monitors BER continuously, and when it becomes higher than 10^{-6}, the receiver sends a feedback signal called the channel quality indicator (CQI) to inform the controller to move to the next transmission rate (lower transmission rate when the BER becomes higher than 10^{-6}).

In our VLC system, we provide three data rates (5, 2.5 and 1.25 Gbps). For example, when the receiver is operating at 5 Gbps and the BER becomes higher than 10^{-6} (i.e. the SNR decrease below 13.54 dB), the receiver will send CQI_1 to inform the controller to reduce the transmission rate to 2.5 Gbps. Then, the receiver measures the SNR, and if the BER is still higher than 10^{-6}, then the receiver will send CQI_2 to inform the controller to further reduce the data rate (i.e. from 2.5 to 1.25 Gbps). Again, if the BER is still higher than 10^{-6}, then the receiver will send CQI_3 to inform the controller to stop transmission, and the communication link is disconnected. However, it is desirable to maintain communication, while a user arbitrarily dims the light source. Therefore, we have introduced a collaborative VLC/IROW system to address this issue (i.e. degradation in the SNR due to dimming will lead to disconnect in the communication link).

It should be noted that the ART has two procedures: down convert and up convert. Down convert is when the controller reduces the data rate due to degradation in the BER (when dimming occurs). Up convert is when the controller increases the data rate (e.g. from 1.25 to 2.5 Gbps) due to maintaining a very low BER at the receiver side (i.e. 10^{-9}). The CQI_4 and CQI_5 signals can be used to inform the controller to increase the data rate from 1.25 to 2.5 Gbps and from 2.5 to 5 Gbps, respectively.

ART is carried out at the start of a one second frame, and if the BER has changed compared to the previous frame's values, then the receiver uses the feedback channel to update the controller.

The ART (down convert) can be applied according to the following steps:

1. The receiver sends (using an infrared beam) a low data rate control feedback signal (CQI) to inform the controller that the BER has become lower than 10^{-6}.

2. The controller decreases the current data rate to the lower service (e.g. 5–2.5 Gbps).

3. The receiver estimates the BER, and if it is still below 10^{-6}, it will send another CQI to inform the controller.

4. The controller further decreases the data rates (e.g. from 2.5 to 1.25 Gbps), and if it receives another CQI from the receiver, the controller will stop the transmission.

A flow chart of the down convert ART is shown in **Figure 7**. To evaluate the performance of the imaging LD-VLC system at different levels of light dimming, the SNR was calculated at 5, 2.5, and 1.25 Gbps. **Figure 8** illustrates the SNR of the VLC system when it was operated at 5 Gbps; the imaging LD-VLC system achieved about a 15.66 dB SNR at the room centre (worst case scenario) when dimming did not exist. However, it can be clearly seen that when the user dims the light by more than 25%, the SNR is decreased, and the BER becomes higher than 10^{-6}. This means that the VLC system cannot operate at this data rate (5 Gbps) when the light is dimmed by more than 25%. **Figures 9** and **10** show the SNR of the imaging LD-VLC system when operating at 2.5 and 1.25 Gbps, respectively. When the imaging LD-VLC is operated at 2.5 Gbps, it is able to maintain a BER of 10^{-6} at a dimming level of up to 50%. On the other hand, the VLC system has the ability to achieve 1.25 Gbps with a BER lower than 10^{-6} at deferent levels of dimming (25, 50 and 75%), as shown in **Figure 10**.

5. Simulation results and discussion of IROW systems

In this section, we evaluate the performance of the proposed support systems in an empty room in the presence of multipath dispersion, receiver noise, back ground noise (light units) and mobility. The results are presented in terms of delay spread and SNR.

5.1. Delay spread

A comparison of the channel delay spreads of our proposed systems is given in **Figure 11**. The receiver moves along the x = 1 m line in the HDIR (with wide FOV and imaging receiver) and BSIR systems, which is considered the worst communication path due to its associated high ISI and path loss. In the CDIR and CDBSIR systems, the receiver moves along x = 2 m, which is considered the worst communication link. The HDIR system with wide FOV receiver shows much more signal delay spread due to the wide receiver FOV (FOV = 90°), which accepts a

http://dx.doi.org/10.5772/intechopen.68474

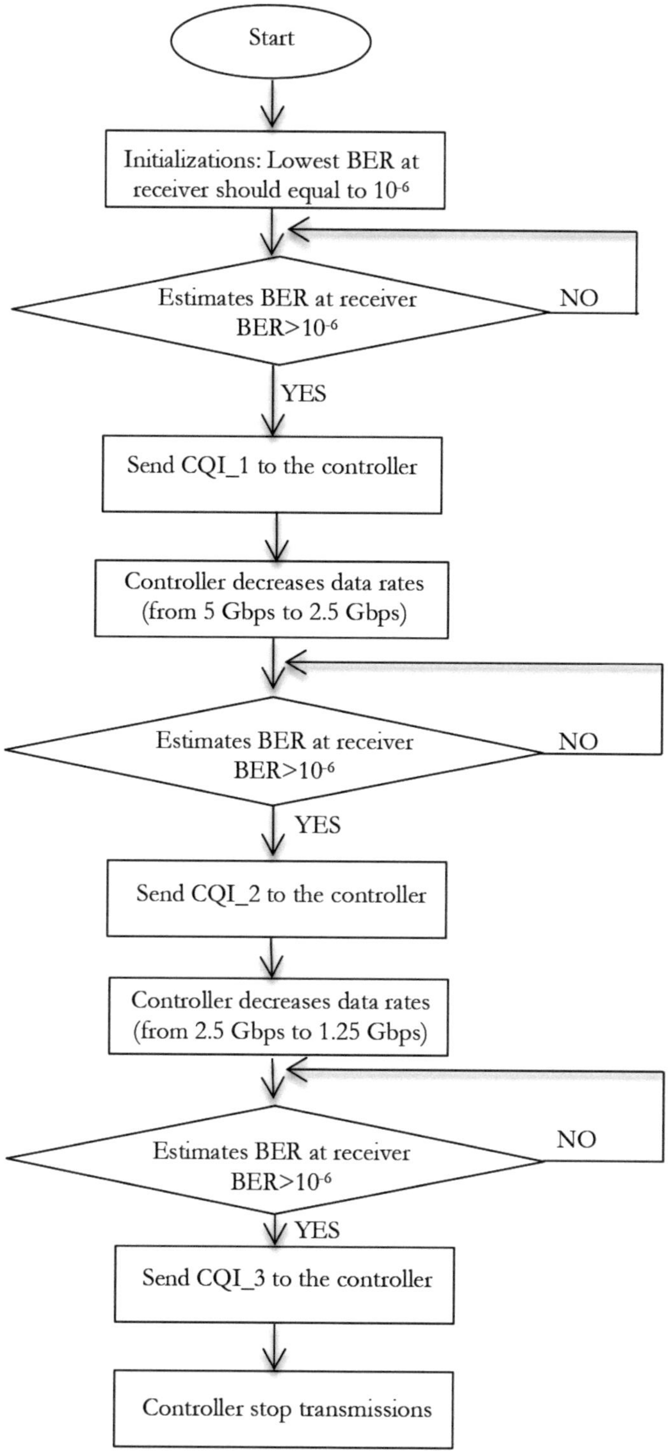

Figure 7. Flow chart of ART (down convert case).

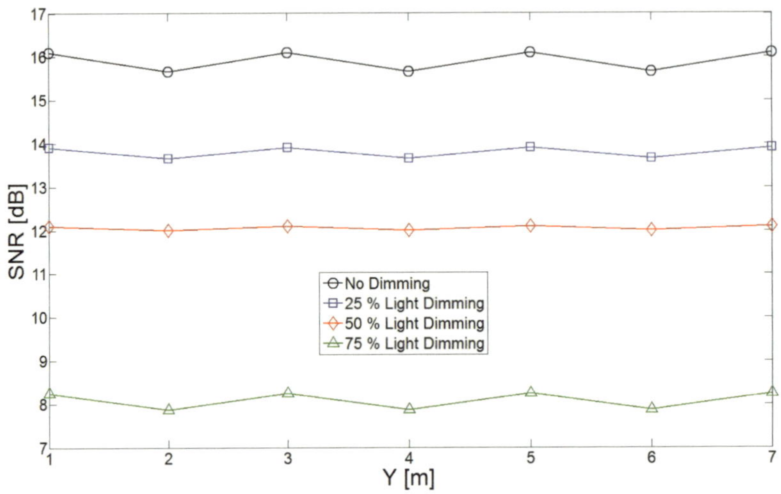

Figure 8. SNR of imaging LD-VLC system operating at 5 Gbps with different levels of dimming (25, 50 and 75%) when receiver moves at x = 2 m along y-axis.

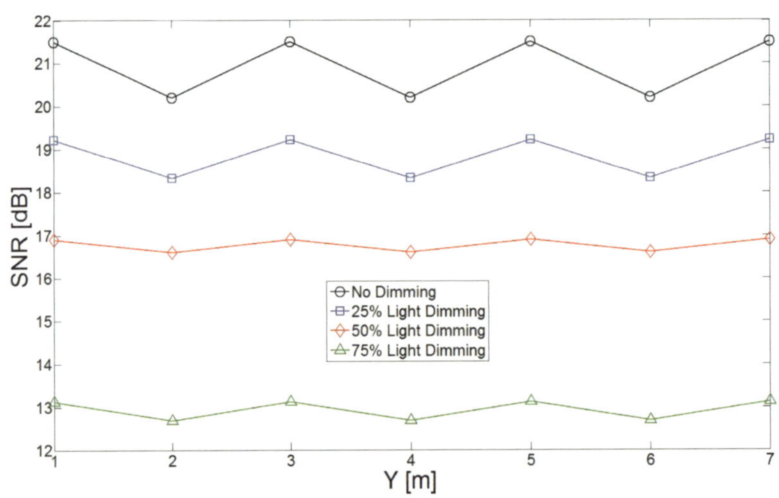

Figure 9. SNR of imaging LD-VLC system operating at 2.5 Gbps with different levels of dimming (25, 50 and 75%) when receiver moves at x = 2 m along y-axis.

wide range of rays with different path lengths from the transmitter to the receiver. In the HDIR system with an imaging receiver, the delay spread results are quoted when the system employs selection combining of the imaging receiver pixels where the pixel with the best SNR (note the SNR expression accounts for delay spread) is selected. The delay spread of our HDIR is reduced from almost 1.55–0.1 ns when an imaging receiver replaces the wide FOV receiver. This is due to the narrow FOV of each pixel, which limits the rays received by using 50 small FOV (about 21°) pixels and selecting the best imaging receiver pixel. The proposed BSIR

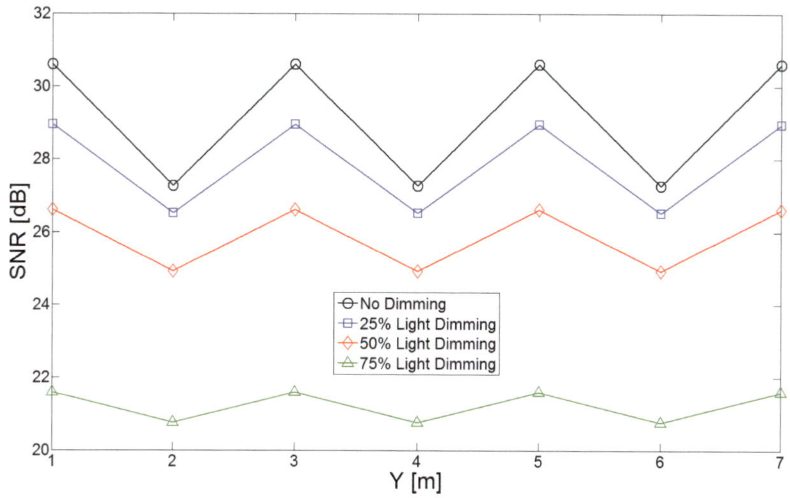

Figure 10. SNR of imaging LD-VLC system operating at 1.25 Gbps with different levels of dimming (25, 50 and 75%) when receiver moves at x = 2 m along y-axis.

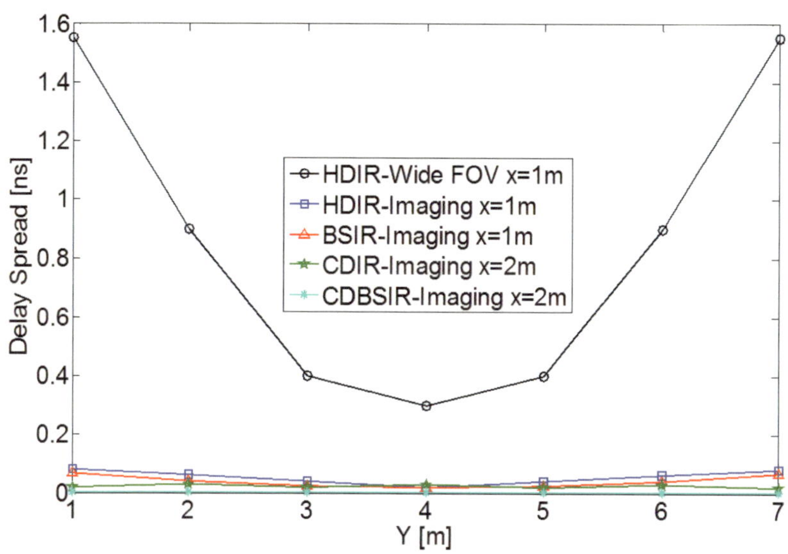

Figure 11. Delay spread of IROW proposed systems.

system coupled with an imaging receiver reduces the delay spread from 1.55 to 0.07 ns. This is attributed to two reasons: firstly, due to the narrow FOVs of each pixel in the imaging receiver, which minimises the number of rays accepted. Secondly, the use of beam steering helps reduce delay spread. It should be noted that steer the IR beam near to the receiver, not only improves the SNR, it is also decreases the delay spread by increasing the received power of LOS component. To further decrease the delay spread, the CDIR and CDBSIR systems are proposed. The CDBSIR system has the lowest delay spread compared to the other systems.

The results show that the CDBSIR system has a lower delay spread than the CDIR system at all the receiver locations considered. The delay spread for the CDIR system is relatively low (0.03 ns in the worst case), and this is attributed to two reasons: firstly, due to the narrow FOVs associated with each pixel in the imaging receiver, and this limitation in the FOV minimises the number of rays accepted. Secondly, the IR transmitter is very close to the receiver (IR sources distributed on the ceiling see **Figure 4**). However, the delay spread can be further reduced (i.e. less than 0.03 ns) by employing beam steering. The CDBSIR system outperforms the CDIR system, as it dramatically decreases the delay spread from 0.03 to 0.003 ns (by a factor of 10) at the room centre. The minimum communication channel bandwidth of the CDBSIR was 29 GHz (where the delay spread is 0.003 ns at points $x = 2$ m, $y = 2$ m, 4 m, 6 m).

5.2. SNR

The SNR evaluation of the proposed backup VLC systems was carried out under the impact of receiver noise, multipath propagation, background noise and mobility. The proposed systems were set to operate at 2.5 and 5 Gbps. **Figure 12** shows the SNR of the proposed systems when operated at 2.5 Gbps. It can be noted that the HDIR system with wide FOV receiver does not have the ability to operate at a high data rate. However, when the imaging receiver is combined with this system, it can perform better than when using the wide FOV receiver. This is because the imaging receiver has ability to select the signals from the optimum pixels that monitor the best received signal during mobility. The imaging receiver uses a large number of detectors with a narrow FOV and small detector area. The HDIR system coupled with an imaging receiver provides around −2.3 dB at $x = 1$ m and $y = 1$ m (worst case), while the HDIR system with wide FOV can only achieve −24 dB at the same location. In OOK, a SNR of 15.6 dB is required to achieve BER of 10^{-9}. A considerable enhancement can be obtained by using the BSIR system, which offers a 33 dB SNR advantage above the HDIR wide FOV receiver at location $x = 1$ m, $y = 1$ m, $z = 1$ m. This enhancement in the SNR is due to the fact that the BSIR system has the ability to steer the IR beam close to the receiver position; hence, the received power will increase. Although improvements were achieved in the BSIR system SNR, a degradation in the SNR is noted when the receiver is on the move (mobile). Therefore, the effect of receiver mobility can be reduced by employing our CDIR system, which is capable of equally covering its environment through the use of a number of IR transmitters distributed on the ceiling.

From **Figure 12**, we can notice that the HDIR (with wide and imaging receiver) and BSIR systems do not able to transfer data higher than 2.5 Gbps; therefore, in **Figure 13**, we only present results for the CDIR and CDBSIR at 5 Gbps. **Figure 13** shows the SNR of the CDIR and CDBSIR systems when operated at 5 Gbps. It can be clearly seen that the CDIR system with imaging receiver does not have the ability to operate at a high data rate. However, when the beam steering technique is combined with this system, it can perform better. The significant improvement in the SNR level is attributed to the ability of the beam steering technique to steer the IR beam towards the receiver location and, thus, increase the power received by the pixels.

It should be noted that only CDBSIR can operate at 5 Gbps. Therefore, we suggested that this system can collaborate with VLC system.

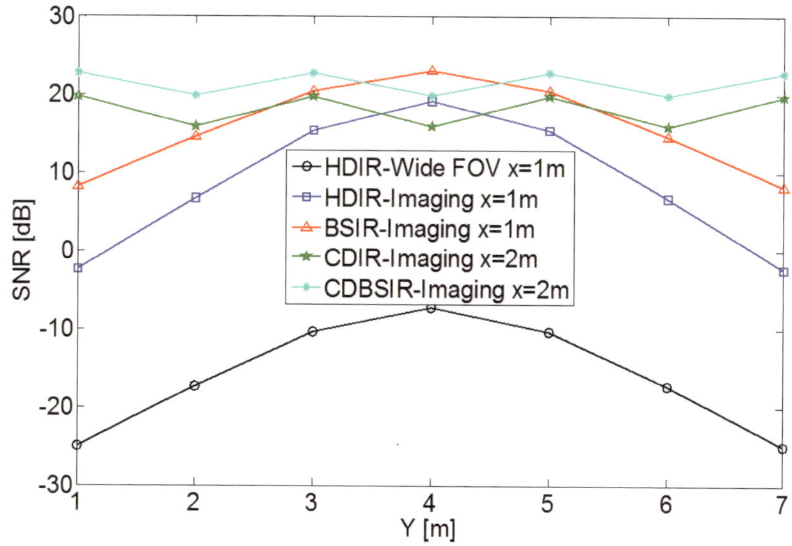

Figure 12. SNR of IROW proposed systems when operating at 2.5 Gbps.

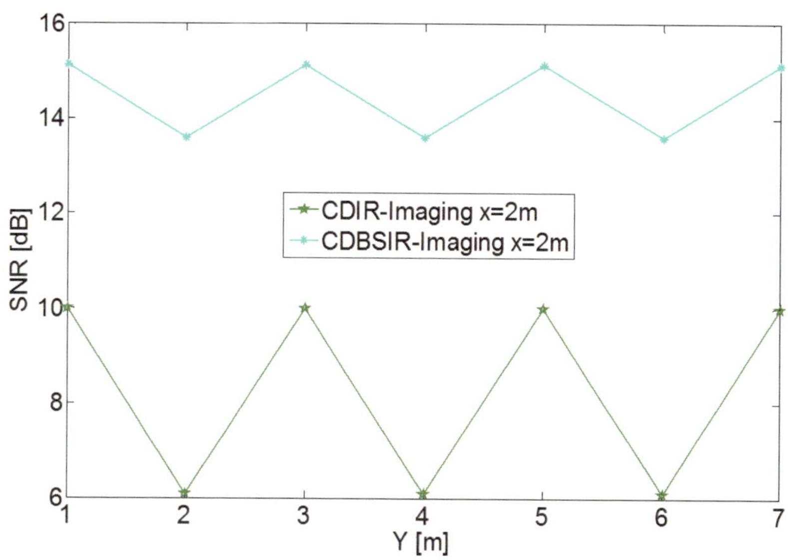

Figure 13. SNR of two systems operating at 5 Gbps when receiver moves at x = 2 m along y-axis.

6. Simulation results and discussion for collaborative VLC/IROW system

Dimming is an important feature of an indoor lighting system where the illumination level can be controlled by the user. One of the main issues in VLC systems is that the light unit should be "ON" all the time to ensure continuous communication. However, the user may dim the light at any time, and this will severally degrade the performance of the VLC system. In this section, collaboration between VLC and IROW systems (CDBSIR) is proposed to support the VLC system when the light is dimmed at different levels (25, 50 and 75%). An ART can be used with the VLC system to manage the reduction in the SNR due to the light dimming and to establish a high-quality communication link under the impact of dimming. To provide a high data rate service for an indoor user under different conditions (with\without dimming), an IROW (CDBSIR) system can be used to support the VLC system. **Figure 14** shows the SNR of the VLC system when the ART is carried out. It can be clearly seen that the data rate diminishes in a very graceful manner when the light is dimmed beyond 50%. However, when employing the CDBSIR system, the achieved data rates at the receiver will be 5 Gbps even though the VLC system is off. It means that the user can dim the lights and maintain a high quality communication service (5 Gbps and beyond). In the case of partial dimming (50 and 75%) in the VLC system, it can achieve 2.5 and 1.25 Gbps, respectively. Therefore, the collaborating system (VLC/IROW) can always achieve higher than the target data rate (5 Gbps). For example, 7.5 Gbps (2.5 Gbps from the VLC system and 5 Gbps from the CDBSIR) can be achieved when the light is dimmed by 50%. In the case of no dimming, 10 Gbps can be achieved by using both systems (VLC and IROW).

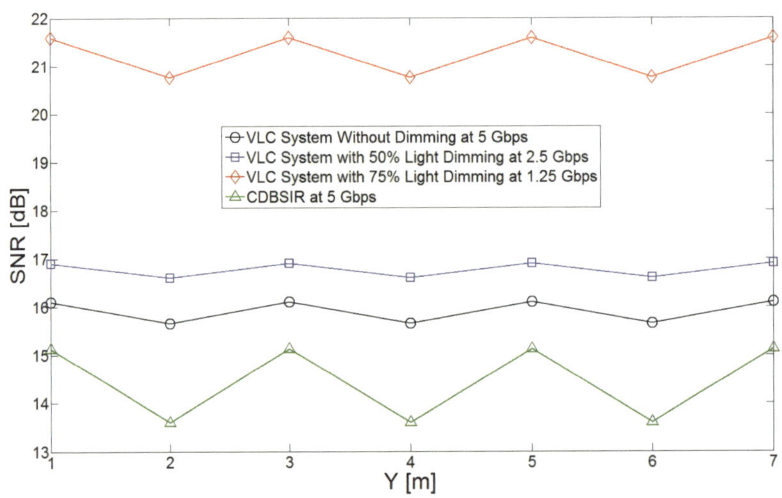

Figure 14. SNR of collaborative systems when receiver moves at x = 2 m along y-axis.

7. Conclusions

In this chapter, we proposed, designed and investigated the concept of a collaborative VLC/IROW system. In addition, we investigated the impact of partial dimming (25, 50 and 75%) on the performance of the VLC system. Moreover, we introduced a novel ART to reduce the effect of the dimming and to create an optimum communication link under the impact of partial dimming.

Five novel IROW systems (HDIR with wide FOV, HDIR with imaging, BSIR, CDIR and CDBSIR) were introduced to support and collaborate with the VLC system in the case of partial dimming or full dimming (i.e. lights off).

Simulation results show that the HDIR and the BSIR systems coupled with an imaging receiver achieved around −2 and 8.2 dB SNR at 2.5 Gbps, respectively. Further improvement in the SNR can be achieved by introducing a new CDIR system and employing more than one IR source distributed on the ceiling (attached to the VLC sources). The simulation results show that the CDIR system can significantly improve the SNR, as well as reduce the delay spread, compared to other systems. A beam steering technique is also proposed to further reduce the delay spread and increase the SNR by steering the IR beam nearer to the receiver at each given location. Simulation results show that the CDBSIR system has the ability to decrease the delay spread of the CDIR system by 90% from 0.03 to 0.003 ns at the room centre ($x = 2$ m and $y = 4$ m), which leads to an increase in the channel bandwidth by a factor of 5.5 from 5.5 to 29 GHz. The simulation results show that the CDBSIR system can significantly improve the SNR. The BER provided by the CDBSIR system is better than 10^{-6} at 5 Gbps in the worst case scenario. Therefore, we used the CDBSIR to collaborate with a VLC system.

Simulation results show that the collaborative VLC/IROW system has the ability to achieve 10 Gbps when dimming does not exist and 6.25 Gbps (5 Gbps from the IROW and 1.25 Gbps from the VLC) in the case of 75% light dimming (worst case scenario). It should be noted that in a collaborative system, the receiver should employ VLC and IR detectors connected through an electronic switching mechanism to control their functions.

Acknowledgements

The authors extend their appreciation to the International Scientific Partnership Program ISPP at King Saud University for funding this research work through ISPP# 0093.

Author details

Ahmed Taha Hussein[1]*, Mohammed T. Alresheedi[2] and Jaafar M. H. Elmirghani[1]

*Address all correspondence to: ahmedtahahussein82@gmail.com and malresheedi@ksu.edu.sa

1 School of Electronic and Electrical Engineering, University of Leeds, United Kingdom

2 Department of Electrical Engineering, King Saud University, Riyadh, Saudi Arabia

References

[1] Pathak PH, Feng X, Hu P, Mohapatra P. Visible light communication, networking, and sensing: A survey, potential and challenges. IEEE Communications Surveys & Tutorials. 2015;**17**(4):2047–2077

[2] Hussein AT, Elmirghani JMH. Mobile multi-gigabit visible light communication system in realistic indoor environment. Journal of Lightwave Technology. 2015;**33**(15):3293–3307

[3] Hussein AT, Elmirghani JMH. High-speed indoor visible light communication system employing laser diodes and angle diversity receivers. 17th International Conference in Transparent Optical Networks (ICTON); 2015. pp. 1–6

[4] Hussein AT, Elmirghani JMH. Performance evaluation of multi-gigabit indoor visible light communication system. The 20th European Conference on Network and Optical Communications, (NOC); 2015. pp. 1–6

[5] Hussein AT, Alresheedi MT, Elmirghani JMH. 20 Gbps mobile indoor visible light communication system employing beam steering and computer generated holograms. Journal of Lightwave Technology. 2015;**33**(24):5242–5260

[6] Alresheedi MT, Elmirghani JMH. Performance evaluation of 5 Gbit/s and 10 Gbit/s mobile optical wireless systems employing beam angle and power adaptation with diversity receivers. IEEE Journal on Selected Areas in Communications. 2011;**29**(6):1328–1340

[7] Alresheedi MT, Elmirghani JMH. 10 Gb/s indoor optical wireless systems employing beam delay, power, and angle adaptation methods with imaging detection. Journal of Lightwave Technology. 2012;**30**(12):1843–1856

[8] Gomez A, Shi K, Quintana C, Sato M, Faulkner G, Thomsen BC. Beyond 100-Gb/s indoor wide field-of-view optical wireless communications. IEEE Photonics Technology Letters. 2015;**27**(1):367–370

[9] Rahaim MB, Vegni AM, Little TDC. A hybrid radio frequency and broadcast visible light communication system. In IEEE GLOBECOM Workshops (GC Wkshps); 2011. pp. 792–796

[10] O'Brien D. Cooperation in optical wireless communications. In Cognitive Wireless Networks: Concepts, Methodologies and Visions Inspiring the Age of Enlightenment of Wireless Communications. Netherlands: Springer; 2007. pp. 623–634

[11] Alsaadi FE, Alhartomi MA, Elmirghani JMH. Fast and efficient adaptation algorithms for multi-gigabit wireless infrared systems. Journal of Lightwave Technology. 2013;**31** (23):3735–3751

[12] Gfeller FR, Bapst UH. Wireless in-house data communication via diffuse infrared radiation. Proceedings of the IEEE. 1979;**67**(11):1474–1486

[13] Haigh PA, Son TT, Bentley E, Ghassemlooy Z, Le Minh H, Chao L. Development of a visible light communications system for optical wireless local area networks. In Proceedings of IEEE Computing Communications and Applications Conference (ComComAp); 2012. pp. 315–355

[14] Cossu G, Khalid AM, Choudhury P, Corsini R, Ciaramella E. 3.4 Gbit/s visible optical wireless transmission based on RGB LED. Optics Express. 2012;**20**(26):501–506

[15] Komine T, Nakagawa M. Fundamental analysis for visible-light communication system using LED lights. IEEE Transactions on Consumer Electronics. 2004;**50**(1):100–107

[16] Hussein AT, Elmirghani JMH. 10 Gbps mobile visible light communication system employing angle diversity, imaging receivers and relay nodes. Journal of Optical Communications and Networking. 2015;**7**(8):718–735

[17] Kahn JM, Barry JR. Wireless infrared communications. Proceedings of the IEEE. 1997;**85**(2):265-298

[18] Al-Ghamdi AG, Elmirghani JMH. Line strip spot-diffusing transmitter configuration for optical wireless systems influenced by background noise and multipath dispersion. IEEE Transactions on Communication. 2004;**52**(1):37–45

[19] Senior JM, Jamro MY. Optical Fiber Communications: Principles and Practice. Pearson Education. University of Hertfordshire, Prentice Hall Pearson Education, USA. 2009; 9780130326812

[20] Leskovar B. Optical receivers for wide band data transmission systems. IEEE Transactions on Nuclear Science.1989;**36**(1):787–793

[21] Pohl V, Jungnickel V, Von Helmolt C. Integrating-sphere diffuser for wireless infrared communication. IEE Proceedings in Optoelectronics. 2000;**147**(4)281–285

[22] Hussein AT, Alresheedi MT Elmirghani JMH. Fast and efficient adaptation techniques for visible light communication systems. Journal of Optical Communications and Networking. 2016;**8**(6):382–397

Chapter 7

Vehicular Visible Light Communications

Bugra Turan and Seyhan Ucar

Additional information is available at the end of the chapter

http://dx.doi.org/10.5772/intechopen.69536

Abstract

Vehicular communications are foreseen to play a key role to increase road safety and realize autonomous driving. In addition to the radio frequency (RF)-based dedicated short range communication (DSRC) and long-term evolution (LTE) communication technologies, vehicular visible light communication (V2LC) is proposed as a complementary solution, utilizing readily deployed vehicle light emitting diode (LED) lights as transmitter with image sensors such as photodetector (PD) and camera as the receivers. V2LC fundamentals including transmitter and receiver characteristics with dimming capabilities are reviewed in this chapter. Depending on the field measurements using off-the-shelf automotive LED light, communication constraints are demonstrated. Moreover, considering the line-of-sight (LoS) characteristics, security aspects of V2LC is compared with the DSRC for a practical vehicle-to-vehicle (V2V) communication scenario. Finally, superiority of V2LC in terms of communication security with the proposed SecVLC method is demonstrated through simulation results.

Keywords: vehicular visible light communications, V2V, VLC dimming, V2LC security

1. Introduction

Next-generation mobility trends such as autonomous driving and ride sharing necessitate various vehicular connectivity schemes. On the other hand, intelligent transportation systems (ITS) harmonized with vehicular communications aim to reduce traffic congestion, accidents, air pollution, energy, and time wastage. Upto date, vehicular communications are expected to provide timely and efficient data dissemination regarding accidents, traffic jams, and road conditions beyond the drivers' knowledge.

Currently, RF-based IEEE 802.11p (DSRC) and LTE are the strongest candidates for V2V and vehicle to everything communications (V2X). However, regarding limited bandwidth and security vulnerabilities of RF-based communications, optical wireless communication is proposed as a complementary technology for vehicular connectivity. Utilizing redundant communication schemes for vehicular communications is expected to increase road safety while supporting safer automated driving applications.

LEDs enable flexible vehicle headlight/taillight design, while providing better illumination, low energy consumption, and longer durability. Hence, LED lights are started to be widely deployed with new production vehicles. Moreover, vehicle LED lights enable creation of various illumination patterns to prevent glare from other road users and illuminate the blind areas better [1]. LED lights illumination requirements and design guidelines are also included in the automotive light regulations [2], which paves the way for more manufacturers to utilize LED in their vehicles. Dimming capability of LED lights is another favorable feature for automotive industry, providing energy efficient vehicular lighting.

Modern vehicles are also equipped with image sensors such as PDs and cameras. PDs are utilized to detect ambient light levels and rain to automatically activate headlights or wiper blades, while the cameras are used for driver assistance systems such as lane keeping assistant, traffic sign recognition, pedestrian detection, and forward collision warning. Hence, usage of the existing vehicle LED lights and image sensors is foreseen to allow low vehicular visible light communication (V2LC) system implementation costs.

Visible light communication (VLC) systems with intensity modulation and direct detection (IM/DD) utilize signal intensity instead of signal phase information. As phase information is prone to distortions for mobility scenarios, sole dependence on signal intensity of IM/DD scheme also makes VLC a promising technology for vehicular communications.

Currently, V2V aims to transmit vehicle position and state information to enhance the road awareness of nearby vehicles. However, with the upcoming autonomous driving features, high-definition real time road maps, vehicle radar data, high-resolution image, and video data from on-board cameras are expected to be exchanged between nearby vehicles. These events driven large size data is required to be conveyed with minimum latency. Furthermore, high mobility requires higher message update rates resulting with dense message generation. In order to provide high data rates with minimum latency, hybrid schemes, utilizing various communication technologies simultaneously, are provisioned to be favorable. It has already been demonstrated that, communication degradation sourced by packet collisions and contention with the usage of single scheme such as DSRC can be avoided with a hybrid scheme employing DSRC and V2LC [3].

Upto date, VLC is reported to achieve multi-Gbit/s data rates for a few meter distances. Compared to DSRC maximum data rate support with 27 Mbps upto 1000 m distances [4], Gbit/s data rates make VLC attractive for high data rate vehicular communications. In addition to higher data rate advantages, with its immunity to malicious jamming with LoS characteristics, VLC is also foreseen to off-load RF networks while providing secure communications for safety critical applications.

In the literature, various studies investigated V2LC applications. In Ref. [5], LEDs are utilized as vehicle ambient lights and dome lights are foreseen to act as VLC transmitters. Authors in

Ref. [6] implemented IEEE 802.15.7 standard to convey information using vehicle LED lights. In Ref. [7], V2LC system that is based on image sensor and high-speed camera receivers is demonstrated.

2. V2LC transmitters

LEDs are used in vehicle headlights, turn signals, taillights, and stop lights. LED arrays with high power form the headlights, fog lights, stop lights, and taillights, whereas single low- or mid-power LED usage is preferred for turn-signals. High-power automotive LEDs draw upto 700 mA current with a typical luminous flux of 200 lm. Low/mid-power off-the-shelf vehicle LEDs turn on with the currents between 60 and 300 mA. Higher current requirements are known to limit the switching capability of LEDs. Furthermore, white LEDs used in the vehicle lighting are usually phosphor-based and have a 3-dB modulation bandwidth in the order of a few megahertz. Phosphor-based white LEDs consist of a blue LED component coated by a phosphor layer. Even though the blue component of LEDs provides upto 20 MHz modulation bandwidth, slow phosphor relaxation time is known as another limitation for the modulation bandwidth of the vehicle LEDs. Thus, higher modulation frequencies of LEDs for V2LC are not considered to be feasible.

Light intensity of an LED source, modulated at an angular frequency of ω is calculated by, [8]

$$I(\omega) = \frac{I(0)}{\sqrt{1 + (\omega\tau_{eff})^2}} \tag{1}$$

where **I(0)** is the light intensity at zero modulation frequency and $\boldsymbol{\tau_{eff}}$ is the effective carrier lifetime. It can be inferred that, ω is also upper bounded with the minimum road illumination requirements as the intensity decreases with the increasing ω. Hence, it is also important to avoid modulation frequencies higher than the 3-dB modulation bandwidth of an automotive LED.

LED light output is linearly proportional to the forward bias current at the linear working region. Driving LEDs at the linear working region is crucial for generating light levels, proportional to the modulation signals. Linear working region of mid-power LED is depicted in **Figure 1a** whereas high-power automotive LED linear working range is shown in **Figure 1b**. **Figure 2** demonstrates the linear working region variation of an automotive LED aimed to be used in headlights with respect to temperature. Even though the illumination degradations linked to nonlinearity of LEDs may not be perceptible with human eye, optical power fluctuations substantially limit the communication performance. Thus, LED nonlinearity should be considered with the selection of modulation and waveforms for V2LC.

V2LC is foreseen to provide high data rates through the usage of multicarrier modulation schemes such as orthogonal frequency-division multiplexing (OFDM). However, high peak-to-average power ratio (PAPR) of OFDM signals necessitates LEDs to work in nonlinear region, resulting optical power efficiency degradation. Hence, PAPR reduction techniques are investigated in Refs. [9, 10] to obtain noticeable bit error rate (BER) performance gain for VLC.

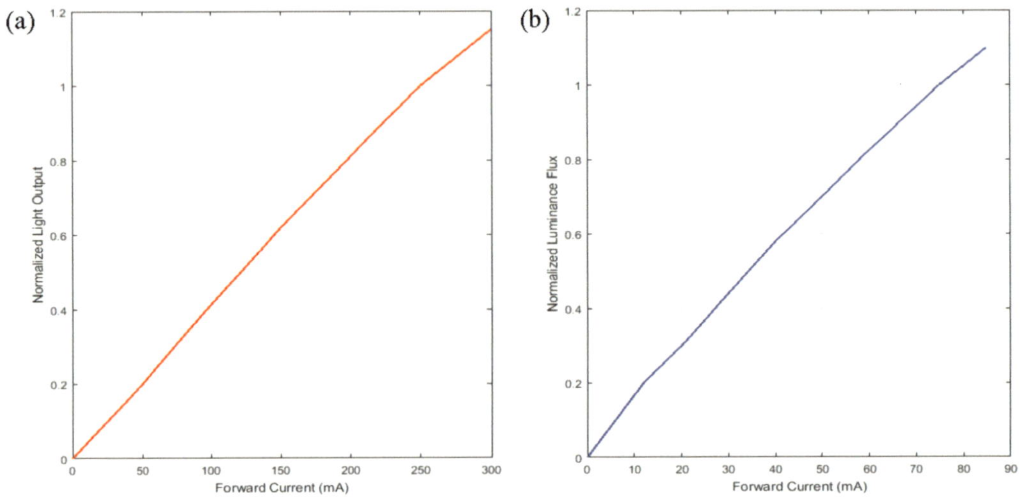

Figure 1. (a) High-power automotive LED bulb; (b) mid power automotive LED bulb linear region.

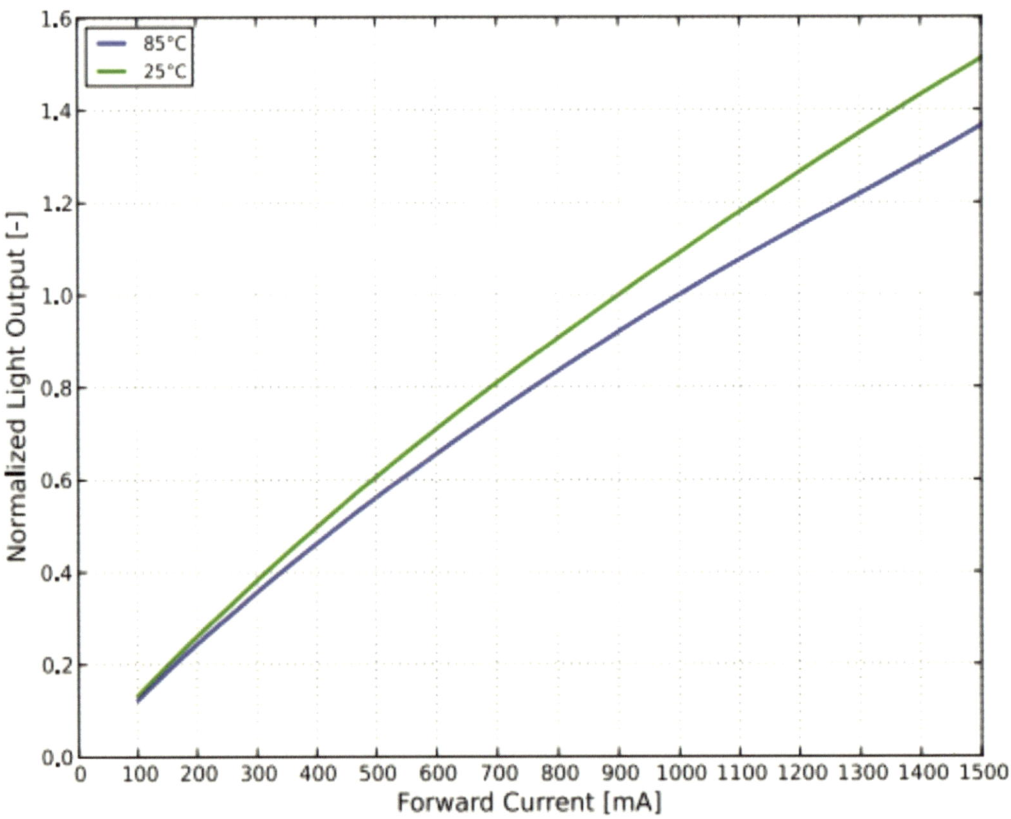

Figure 2. Automotive LED linear working region variation with temperature.

LED radiation patterns define the relative optical power strength from the light source. Almost every single LED chip is designed to emit Lambertian pattern. However, refractions and internal reflections inside the encapsulating housing describe the final radiation pattern of LEDs. As vehicle LEDs are encapsulated into housing with reflectors and lenses, pure Lambertian radiation pattern assumption is not practical for V2LC applications employing LED arrays such as headlights and taillights. However, certain vehicle lights, utilizing single LED, can be approximated to Lambertian radiation model as detailed in Section 2.2.

2.1. Lambertian model

Single LED sources are generally designed to have Lambertian beam distribution where the spatial luminous intensity distribution is a cosine function of Lambertian order and half-intensity beam angle.

Optical channel gain of the Lambertian radiation pattern is defined by,

$$H(0) = \begin{bmatrix} \frac{(m+1)A_{pd}}{2\pi d^{\gamma}} cos^{m}(\varphi)T_{s}(\theta)g(\theta)\cos(\theta), \quad 0 \leq \theta \leq \theta_{c} \\ 0, \; elsewhere \end{bmatrix} \tag{2}$$

where d is the inter-vehicular distance, $\boldsymbol{\varphi}$ is the irradiance angle, $\boldsymbol{\theta}$ is the incidence angle, $\boldsymbol{\theta_c}$ is the PD field of view (FOV), $\mathbf{A_{pd}}$ is the active receiver area of PD, γ is the path loss exponent, $\mathbf{T_s(\theta)}$ is the filter gain, $\mathbf{g(\theta)}$ is the gain of an optical concentrator calculated by,

$$g(\theta) = \begin{bmatrix} \frac{n^2}{sin^2(\theta)}, \; if|\theta| \leq \theta_c \\ 0, \; if \; |\theta| \geq \theta_c \end{bmatrix} \tag{3}$$

in which n is the internal refractive index of PD and m is the order of Lambertian model specifying transmitter directivity and obtained by,

$$m = -\frac{\ln 2}{\ln(\cos \varnothing)} \tag{4}$$

where $\varnothing$ is the half-intensity beam angle of an LED. The coverage range and radiation pattern of single LED light is affected by the half-intensity beam angle $\varnothing$ such that narrower $\varnothing$ increases the illumination range. The average received optical power $\mathbf{P_r}$ is computed by;

$$P_r = H(0)P_t \tag{5}$$

Half-intensity beam angle of a single LED light can be accessed through data sheets provided by manufacturer. Moreover, PD and optical receiver-related parameters such as $\mathbf{A_{pd}}$, $\mathbf{T_s(\theta)}$, $\mathbf{g(\theta)}$are also accessible through product specifications. Irradiance ($\boldsymbol{\varphi}$) and incidence angle ($\boldsymbol{\theta}$) values can be calculated through the relative locations of the transmitter and receivers on the vehicle.

2.2. Approximated Lambertian model

LEDs are encapsulated in automotive light housing with collimating and diffusing optics including lenses and reflectors. Thus, headlight, taillight, or stop light can be considered as a single radiation source. Furthermore, the half-intensity beam angle of a single LED is no longer applicable to Eq. (1) in order to obtain the optical channel gain. Additionally, for mobile scenarios, path loss component (γ) is directly related to half-intensity beam angle due to rapid irradiance and incidence angle variations. Single LED light without a lens, emits optical power within a smaller angle. Thus, path loss exponent is more prone to fluctuations with vehicle mobility. On the other hand, vehicle LED lights employing either a lens or reflector, provide increased irradiance angle when compared to a single LED and experience less fluctuations sourced by the mobility.

Half-intensity beam angle of a vehicle light is not specified by manufacturers and can be estimated through measurements. In order to model the VLC channel practically, path loss exponent also needs to be estimated through received optical power measurements for various light conditions.

Linear least squares algorithm is used in Ref. [11] to estimate the half-intensity beam angle and path loss exponent. In **Figure 3a**, optically received power comparison between ideal Lambertian radiation pattern and the measured power levels from an automotive LED fog light is demonstrated. **Figure 3b** depicts the estimated Lambertian radiation pattern whereas the measured radiation pattern is shown in **Figure 3c**. Half-intensity beam angle is estimated to be 50.66° while the path loss exponent is calculated as 1.8139 according to the nighttime static measurements. It can be observed that, decremented patterns for both the estimated Lambertian model and the actual measured fog light are consistent. However, the actual measurements indicate the increased intensity due to vehicle light optics.

Vehicle light irradiance angle upper limits are defined with the automotive lights regulation [12]. According to the regulation, vehicle fog lights are expected to illuminate upto 26° . viewing angle. In **Figure 4**, regulation compliant measurement resulted from an off-the-shelf automotive fog light is depicted. It can be concluded that, received power decreases with both the increasing distance and angles. Power degradation depending on the incidence angle should be highlighted, when compared to RF-based vehicular communications. However,

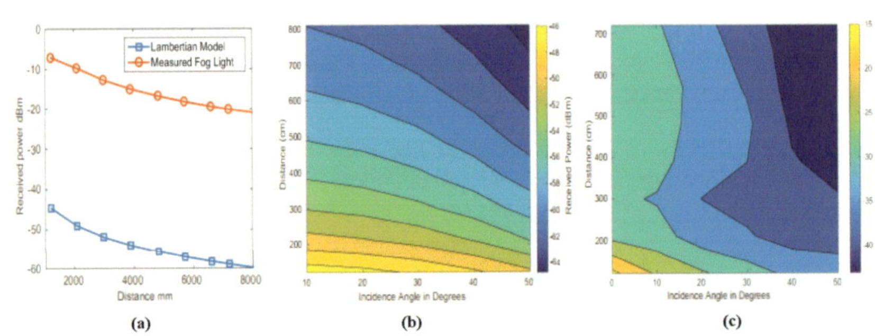

Figure 3. (a) Ideal Lambertian vs. measured model radiation pattern; (b) estimated Lambertian radiation pattern; and (c) measured radiation pattern.

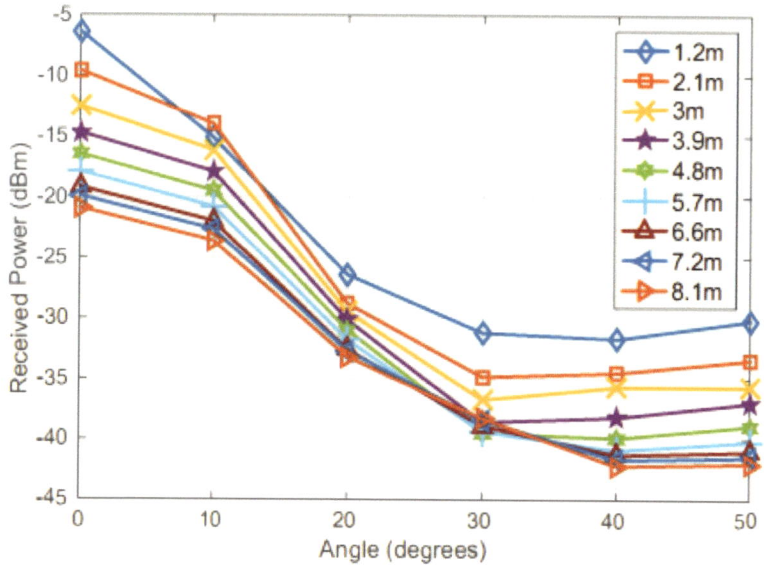

Figure 4. Off-the-shelf automotive fog light received power measurements.

apart from the traditional vehicle lights, newly developed custom pattern generating LED lights such as matrix LED automotive lights are foreseen to provide a flat illumination and stable optical power, similar to RF radiation [13].

3. V2LC receivers

PDs, PD arrays, or cameras can be employed as V2LC receivers. Depending on the sensors FOV, location and lens selection plays a key role to realize practical V2LC applications. Sunlight and artificial background lights are also supposed to be considered at the receiver side. Direct exposure to sun light causes either saturation or excess shot noise at the image sensors resulting with inability to detect intensity-modulated signals. In Ref. [14], direct current (DC), suppressing front-end circuit usage, is proposed to suppress noise sourced by sunlight. Artificial light sources such as advertising boards and traffic signals are usually operated at the 60 Hz AC voltage and its harmonics. Hence, modulating V2LC LEDs in the order of at least a few hundred kHz to a few MHz is beneficial to minimize artificial light noise effects.

PDs target high-rate VLC, whereas cameras are foreseen to support low-rate VLC with positioning capabilities. Current vehicles with automatic headlight and rain sensors are already equipped with PDs. On the other hand, the number of vehicles with cameras is increasing to enable features such as forward collision warning, pedestrian detection, traffic sign detection, and lane keeping. Both camera and PD sensors are located in the middle of the windshield above the rear-view mirror. Even though the already deployed image sensors can be utilized for V2LC, experiments in Ref. [15] located the sensors, around the headlights and taillights to evaluate the performance dependence to the multipath reflections from the road.

Responsivity of PDs is a measure of the sensitivity of PD to the light input. It is defined as the photocurrent ratio (**I_p**) to the incident light power (P) at a given wavelength by;

$$R_p = \frac{I_p}{P} \tag{6}$$

Figure 5 depicts typical spectral responsivity of a low-cost silicon PD, which can be employed in vehicles as a V2LC receiver.

Fraction of the incident photons contributing to photocurrent is defined as the quantum efficiency (QE) of a PD, given as,

$$QE = \frac{R_{p\ \text{observed}}}{R_{p\ \text{ideal}}} \tag{7}$$

QE is known as the capability of a PD to convert light energy to electrical energy. In **Figure 6**, QE variations sourced by operating temperature are demonstrated.

White LED's spectral response (**Figure 7a**) depicts that the blue component is dominant, whereas for the tail and stop lights, red component in the wavelength interval of 600–640 nm is dominant (**Figure 7b**). Hence, it can be concluded that different PDs should be employed for the front and rear of the vehicles as the front PDs are expected to capture red LED taillights whereas the rear PDs are foreseen to capture white LED headlights.

V2LC systems can also utilize readily deployed vehicle on-board cameras. Camera usage effectiveness heavily depends on image processing or computer vision techniques for emitter

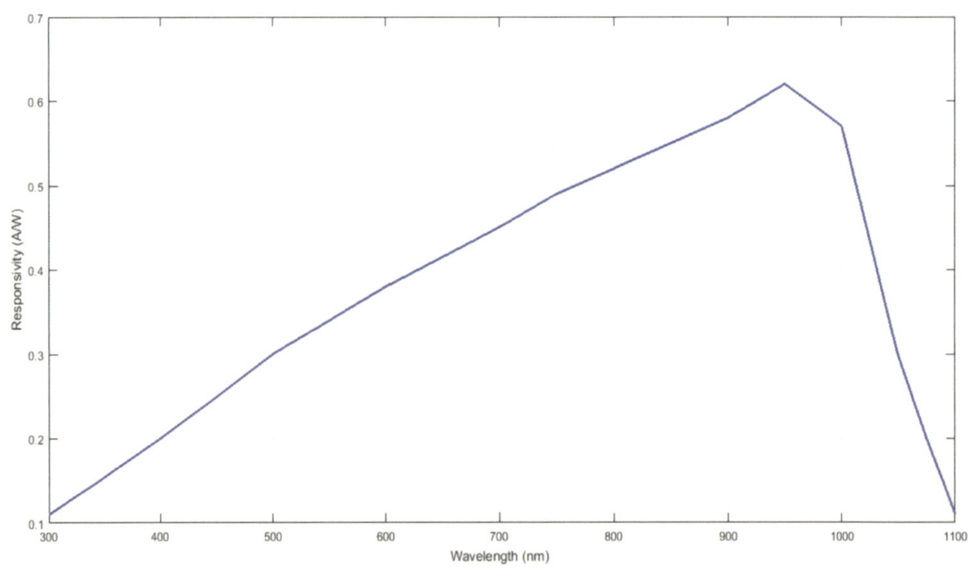

Figure 5. Typical spectral responsivity of Si-PD (Thorlabs PDA100A).

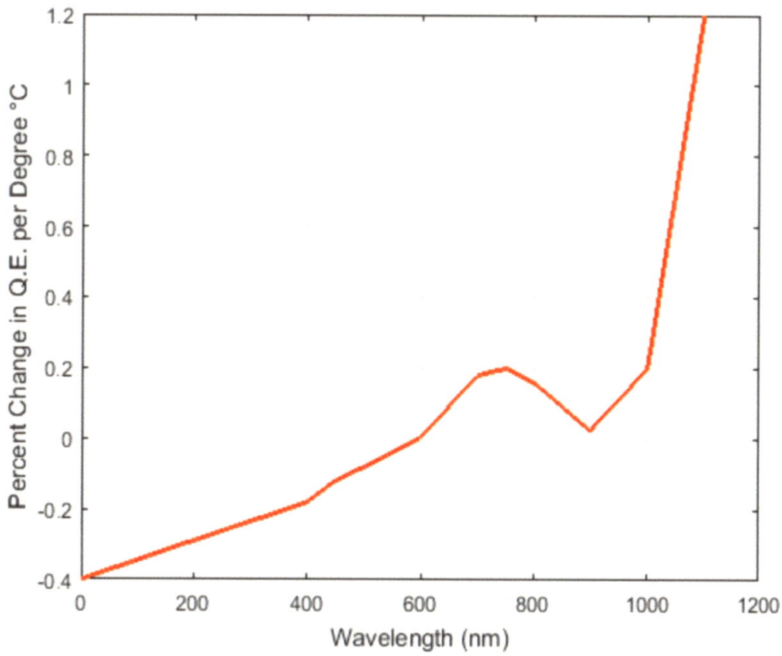

Figure 6. Temperature dependence of quantum efficiency.

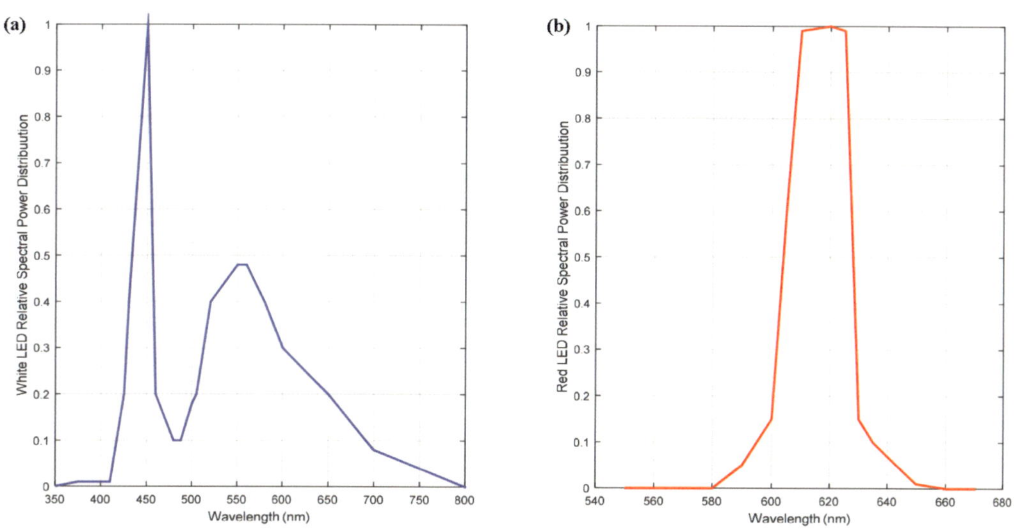

Figure 7. (a) White automotive LED spectral response; (b) red automotive LED spectral response.

finding and tracking. On-off-keying (OOK) and rolling shutter effect-based modulation schemes are proposed to be applicable for V2LC applications.

High speed camera with 1000 frames/s (fps) is shown to provide road to vehicle VLC upto 60 m distance at 4 kbps rate with a BER of. 10^{-3} [16].

Photodetector	Wavelength (nm)	Responsivity (A/W)
Silicon PN	550–850	0.41–0.7
Silicon PIN	850–950	0.6–0.8
InGaAs PIN	1310–1550	0.85
InGaAs APD	1310–1550	0.8

Table 1. Typical photodetector characteristics.

Using camera with customized image sensor, 10 Mbps data rate is achieved in Ref. [17], whereas 45 Mbps without bit errors is achieved in Ref. [18].

Currently, off-the-shelf image sensors utilized in the vehicles have extended dynamic ranges upto 120 dB with approximately 30 fps capture rates. Therefore, off-the-shelf automotive cameras provide superior detection capabilities for various light conditions but the data rate is limited to 15 bit/s (bps) or less to satisfy the Nyquist frequency requirement. Furthermore, additional time needed for image processing to detect and track the LED transmitter, should carefully be considered to fulfill communication latency requirements (**Table 1**).

4. V2LC channel

4.1. V2LC dimming support

It is possible to dim LEDs to an arbitrary level depending on the application requirement in order to save energy. Dimming capability of LEDs makes them also favorable for vehicle lighting. Depending on day light conditions, vehicle lights are expected to change their brightness to prevent glare and increase road safety while providing energy efficiency with the dimmed LEDs. Furthermore, dimming capability of vehicle LED lights gains attention in terms of life span, as dimmed LEDs require less current; hence produce less heat, which extends lifetime of an LED bulb. Dimming is provided via changing the forward bias current through the LED. In terms of communication, dimming has crucial effect on signal-to-noise ratio (SNR), with the achievable data rate and BER. Thus, detailed analysis of dimming effects on V2LC and proper dimming methods or protocols should be developed to provide the right trade-off between illumination and communication. Analysis of dimming functionality and efficient dimming techniques in V2LC systems will contribute to the safety and allow the vehicular system to have full control over the lighting output.

V2LC systems are foreseen to be designed in a fashion where data transfer is maintained while the LEDs are dimmed. Two different dimming schemes are proposed at the IEEE 802.15.7 standard [19]. Using OOK modulation, additional time slots for the on and off times are considered, where these time slots decrease the transfer data rate and keep the maximum communication distance constant. Furthermore, on and off level light intensities can be redefined to achieve dimming with OOK modulation where the data rate is kept constant with dimming. Adding compensation symbols to match target dim level is classified as the time domain approach whereas changing intensity levels or frequencies of message symbol occurrences is regarded as intensity domain approach [20] in the VLC dimming literature.

Intensity domain approach is utilized with the DC bias scaling. In DC bias scaling, to decrease brightness up to 25%, OFF symbol DC bias level is increased and ON symbol DC bias level remains constant. For further dimming upto 75% of the regular illumination, ON symbol bias level decreases while keeping OFF symbol DC bias level constant.

Compensation symbol adding decreases bandwidth usage, hence limits high data rate communications. Therefore, intensity domain approach provides more advantage in terms of high data rates.

Figure 8a depicts the received power with dimming support and **Figure 8b** denotes the DPDR performance for four dimming levels of a VLC hardware adapted for V2LC using off-the-shelf automotive LED fog light. Intensity domain approach is utilized where dimming level 0 denotes the minimum intensity and the dimming level 9 denotes the regular brightness level. Outdoor measurements, emulating platoon distances upto 5 m are conducted at static scenarios where the transmitter and receiver units are aligned at 36 cm height. 100 packets are conveyed to evaluate the data packet delivery ratio (DPDR) with respect to dimming levels. Dim level is shown to have crucial effect on communications, as 100% DPDR is achieved with normal brightness at 10 m distance while no packet delivery is possible at 5 m distance with dimming. It is eminent to note that, with the intensity domain dimming, even though the target data rates are preserved, communication reliability decreases as the decreased ON signal intensity levels lower the detectability of ON-OFF signal differences. However, utilizing time domain dimming approaches, increased reliability can be expected as the optical power levels of the message signal remains constant and the power differences between ON and OFF signals stay more detectable at the receiver.

Experiment results indicated that, knowing the distance from the receiver, vehicle LED dim levels could be adjusted to ensure reliable data dissemination while preventing sudden glare for safety reasons. High mobility of vehicles requires adaptive dimming schemes to guarantee communication reliability with road safety and energy efficiency.

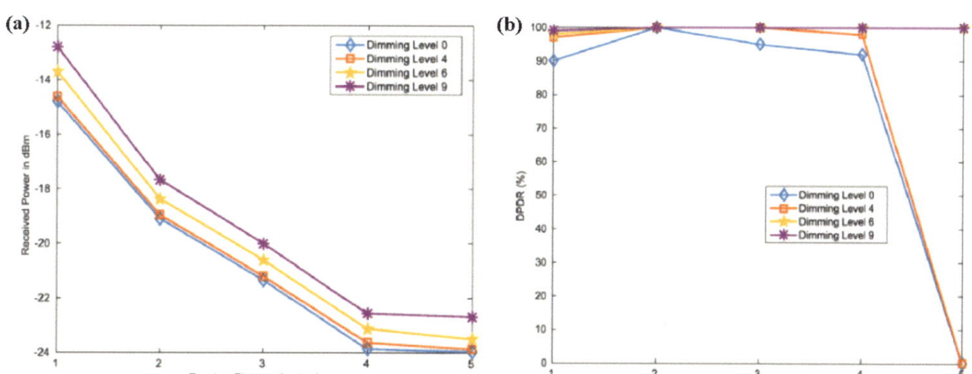

Figure 8. (a) Received power with dimming support; (b) DPDR for four dimming levels.

Considering the variable pulse-position modulation (VPPM), pulse width is varied to control dimming. However, changing pulse width, data transfer rate remains constant whereas the maximum communication range decreases. Vehicle LED lights are already employing pulse-width modulation (PWM) scheme to prolong the life cycle of LEDs and obtain energy efficiency. Hence, combining pulse-position modulation (PPM) with PWM, VPPM can be deemed as a favorable modulation technique to utilize dimming property of LEDs. Moreover, for the applications targeting constant following distances such as platoons, as the maximum communication distance is already limited, keeping transfer rate constant becomes more favorable. Dimming support for OOK and PPM are well defined and can effectively be utilized. However, inter-symbol interference with the OOK and PPM limits the achievable spectral efficiency and Gbit/s data rates.

LEDs dynamic range is crucial to prevent nonlinear distortion sourced by the clipping of OFDM signals. Hence, linear working region of vehicle LEDs plays a key role for efficient dimming support of V2LC devices.

5. Secure vehicular communications through V2LC

One promising application area of V2LC is vehicular platoon where a group of cooperative adaptive cruise control (CACC) vehicles kept in close proximity through DSRC. In the vehicular platoon, inter-vehicular space gap is less than 15 m at vehicle speeds less than 100 km/h [21]. On the other hand, VLC communication range has been demonstrated to be 100 m for headlights and 30 m for taillights [15]. Moreover, the light directivity and impermeability of the optical signal through vehicles and obstacles provide more secure communication than DSRC by limiting the transmission area. This limited transmission area restricts the availability of the data to the attackers, while still allowing communication in the vehicular setting. However, utilizing only VLC in vehicle platoon may degrade communication, since VLC is sensitive to environmental effects, i.e., fog, and might have short-term unreachability due to the increase in the inter-vehicle distance and/or LoS on a curvy road. Thus, IEEE 802.11p and VLC hybrid architectures are proposed to provide redundancy for better reliability in vehicular platoons [21–24].

Recently, many researchers are aggressively investigating V2LC for different purposes such as channel characteristics [25, 26], requirements [9, 27–29]. Proposed V2LC schemes were studied either experimentally [5, 9, 15, 27] or via computer-based simulations using the Lambertian property of LEDs [26, 28]. Most of the academic research on V2LC has targeted achievement of high data rates by using advanced modulation schemes [20, 30, 31]. Since the history likes to repeat itself, a common mistake in novel applications was to underestimate the security issues [24]. Currently, the V2LC industry is on the same path again where the V2LC applications have directed researchers focus away from the security issues. However, one of the features that make the V2LC alternative to the DSRC is security.

In this section, we outline the security implications of V2LC and present the possible approaches to secure the vehicular communication. Finally, a case study where a secure V2LC

http://dx.doi.org/10.5772/intechopen.69536

protocol is presented to enable efficient and secure data sharing among vehicles via the usage of VLC directionality property.

5.1. Securing vehicular visible light communication

Vehicular network security attacks can be categorized into three parts as demonstrated in **Figure 9**. In system level attack, vehicle hardware/software or security certificate is targeted by either malicious insider in manufacturing level or by an outsider. Via altering critical vehicle parts such as sensors, a malfunctioning system is intended. Even though the vehicular communication channel is secure, vehicle generates faulty information that has a catastrophic effect on safety. Furthermore, application and network layer attacks targeting vehicular communication may have substantial effects on vehicular communications security. Application layer attacks can be summarized as; spoofing, where an adversary imitates another vehicle within the transmission range and injects wrong information into the network; replay attack, where the disseminated packet is stored by adversary and replayed at a later time to ruin vehicular security, and eavesdropping, where adversary secretly collects the transmitted packet within its transmission range and processes the packet information. Compared to application layer, network layer attacks aim multiuser applications where in denial-of-service (DoS) and distributed DoS communication medium is congested via redundant transmissions. In RF radio jamming, particular geographical locations are attacked via DoS and vehicles in the area are disturbed. Research findings on various types of attacks using the actual vehicle have been presented in the past 5 years [32–35].

In real-world V2LC deployment, the communication needs to be protected against predefined security attacks. Moreover, to ensure the security in V2LC, three requirements; authentication,

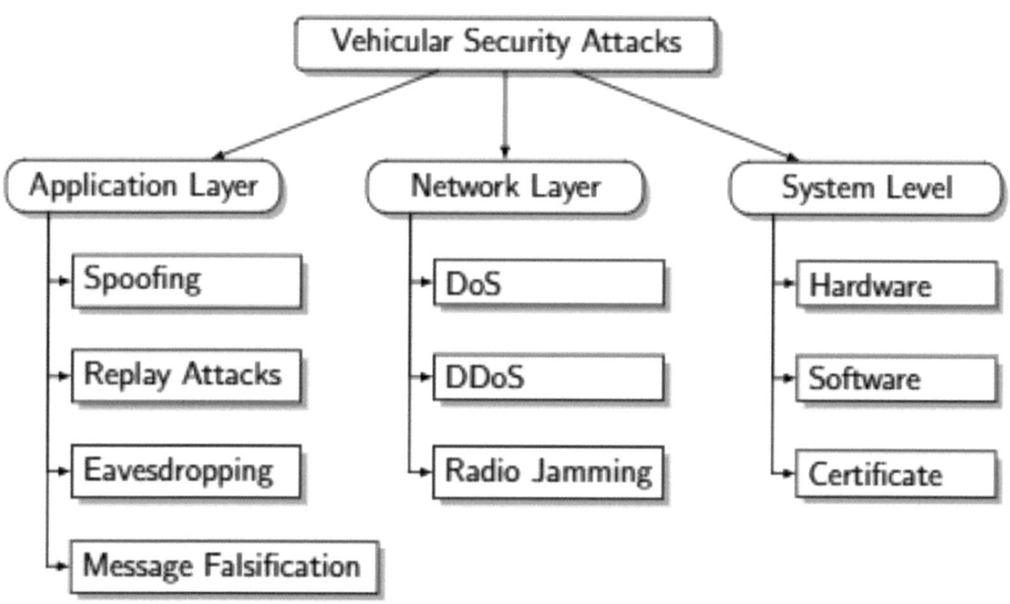

Figure 9. Categorization of vehicular network security attacks.

confidentiality, and integrity must be satisfied. Authentication confirms the identity of the vehicle and gives permission to authorized vehicles to access the communication medium. Confidentiality ensures only the participating vehicles decode the content of messages. Integrity, on the other hand, confirms that the transmitted data is not modified during over the air transmission. To enable the authentication, confidentiality and integrity for V2LC networks, physical layer protection, stenographic protection, and cryptographic key generation/management security methodologies are foreseen to be exploited.

5.1.1. Stenographic protection

Steganography relies on keeping a secret data within the transmitted message. It can be combined with VLC where a secret message can be placed in existing communication. In V2LC, two endpoints share a secret that defines the camouflage of the message within the communication. Despite the steganography ensures the confidentiality, it suffers from the lack of authentication and integrity. However, hiding a secret light beam in V2LC is worth investigation for devices that do not have resources to run complex operations.

5.1.2. Cryptographic key generation and management

In cryptographic key generation/management, vehicles use secret keys to secure the VLC based on either asymmetric cryptography [36] or symmetric cryptography [37, 38]. In asymmetric cryptography, sender and receiver establish a secret key by using pairs of keys. The public key can be disseminated publicly, while the private key is only known to the owner. Sender and receivers agree upon a secret key using a key establishment protocol periodically. On the other hand, in symmetric cryptography, a shared key between two or more vehicles is used to maintain a private information link. These secret or shared keys are then used in the encryption and decryption of the message at the sender and receiver, respectively.

Allowing access to the secret key by two or more vehicles makes symmetric key encryption vulnerable to security attacks. As an alternative, asymmetric cryptography has been recently proposed for vehicles where group key is cooperatively established via RF communication [33]. The shared group key is then used to secure the communication among vehicles.

Unlike steganography, cryptography ensures the confidentiality via encryption/decryption, integrity with hashing, and authentication by using message authentication codes. The resilience of VLC against jamming attack makes it possible to propose a hybrid architecture where VLC is used for key generation and the generated key is used for securing the both VLC and RF communications.

5.2. Case study

In the case study, a secure light communication protocol, namely SecVLC, is proposed. The unique features of SecVLC are as follows:

1. It uses the directionality property of VLC to ensure only target vehicles participate in the communication.

2. It utilizes the full-duplex communication where infrared (IR) is the outgoing link to share a secret key and VLC is the incoming link to receive encrypted vehicle data.
3. It operates with keys generation and share mechanism that is used for the data encryption and decryption where data packets cannot be decrypted without generated keys.

SecVLC stands for cryptographic key generation/management where vehicles use IR to share a secret key. After receiving the secret key from the destination, the source encrypts the data packet based on Advanced Encryption Standard (AES) and transmits the encrypted packet via light beams in VLC.

SecVLC is implemented in Java on top of Li-1st transceiver software [39] that is integrated in Keyczar [40] key generation toolkit. Li-1st is the first commercial product of VLC that is manufactured by pureLifi Ltd [41]. It provides an opportunity to rapidly develop and test VLC applications that utilize commercial LED infrastructures. Li-1st consists of transmitter unit (Tx) and PD-based receiver unit (Rx). Tx is attached to two symmetrical LED fog lights [42] where automotive fog lights are preferred due to their wide and flat illumination pattern to minimize reflection by fog. On the other hand, Keyczar is an open source toolkit developed by Google for key generation. Experimental setup parameters are depicted in **Table 2**.

Two Vishay [43] high-speed infrared emitting diodes are utilized as IR transmitter and IR receiver for sharing the secret key between source and destination. Both Tx and Rx are connected to computers for evaluating communication performance. In order to compare the security vulnerabilities of communication medium, scenarios, where vehicles use DSRC and visible light data transmission, namely VLC, are evaluated. The DSRC communication scenario is simulated with the convoy driving implemented simulator, VEhicular NeTwork Open Simulator (VENTOS) [44]. On the other hand, VLC and SecVLC experiments are performed in an outdoor environment as shown in **Figure 10** to take into account the reflections from vehicles and road. Nighttime outdoor measurements are executed to compensate shot noise, sourced by diurnal variations. Our experiment emulates the scenarios that are the front of following vehicle-disseminating commands (i.e., mission orders, mission plan, etc.) with LED

Parameters	Value
LED fog lights ground height	36 cm
LED fog lights separation distance	150 cm
Inter-vehicular distance	2–6 m
Data packet size	100 bytes
Li-first modulation	Pulse-amplitude modulation
Li-1st error correction	Reed–Solomon
Li-1st data rate	5 Mbps
Secret key packet rate	4 bytes
Vishay IR half intensity	18°

Table 2. Experimental setup parameters.

Figure 10. V2LC experimental setup.

fog lights to the rear of leading vehicle proceeding on a curved path. **Table 2** lists the experimental system parameters.

Performance evaluation of SecVLC focuses on the security analysis of SecVLC where the system with a malicious insider is investigated. The malicious insider is a vehicle that is positioned on the road with constant mobility. For each experiment, 100 data packets are sent over the network and malicious insider tries to extract the data content. In security analysis of SecVLC, malicious vehicle's data decoding ratio is analyzed. Data decoding ratio is defined as the ratio of the number of successfully plain text converted data packets to the total number of transmitted data

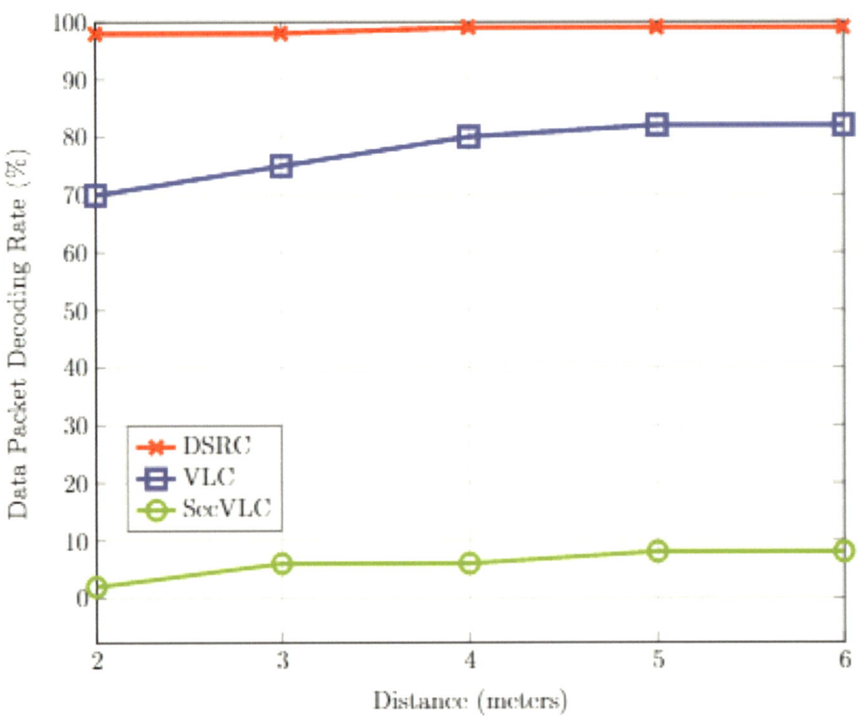

Figure 11. Data packet decoding rate comparison.

packets. In this scenario, the malicious vehicle receives the data packets and tries to decode the data for subsequent processes such as stealing the vehicle identity information.

Figure 11 demonstrates that adversary vehicle can receive the data packet in both DSRC and VLC scenarios with minimum 70% data packet decoding ratio. In the DSRC, adversary vehicle overhears the channel if it is located in the transmission range (300 m) of transmitting vehicle. On the other hand, VLC limits the adversary data reception due to its directional transmission. However, adversary vehicle still receives the data if it is positioned in headlight coverage. Compared to DSRC and VLC, SecVLC encrypts the data packet and data content can only be decrypted with the secret key. Even if the adversary vehicle overhears the channel, it can only receive plain text control packets transmitted in the initialization phase of the protocol.

Experimental evaluation of SecVLC demonstrates its suitability for securing light-based vehicular communication. In the security analysis of SecVLC, it is observed that despite VLC limits the data reception due to its directional transmission, it is still possible to receive and decode the data packet if the adversary is located inside the light coverage. Furthermore, secret key enabled SecVLC prevents data packet decoding of adversary vehicle even though it is captured successfully.

6. Conclusion

In this chapter, V2LC key features such as automotive LED characteristics, PD features, dimming functionality application to vehicle LEDs, and vehicular communication security enhancement through complementary usage of V2LC are discussed. Apart from indoor VLC, V2LC requires high power LEDs with collimation and diffuse optics to fulfill long-range road illumination for all weather and ambient light conditions. High power LEDs limited switching capability, white LED usage and minimum illumination requirements for road safety upper bound the modulation frequencies.

PD or image sensor selection also plays a key role to realize V2LC practical applications. Receivers located in front of the vehicle are expected to capture taillights in red color, whereas the rear optical receiver sensors are foreseen to capture communication signals transmitted through white LEDs. Furthermore, image sensor or high-speed camera usage is also practiced with various experiments, and data rates above RF-based DSRC communications are achieved. Image sensor usage is also deemed favorable due to positioning capabilities of optical image sensors.

LEDs dimming capability enables energy savings prolong LED life and adaptive illumination for safer traffic. However, considering power and illumination limitations of outdoor environment, time domain dimming schemes providing more SNR at the receiver are regarded to be practical, despite the data rate limitations.

Hybrid usage of V2LC and RF-based DSRC enables enhanced security for vehicular communications. As both technologies can complement each other in terms of data rates, directional communications and range extension, exploiting both technologies for vehicular networks, are demonstrated to be practical in terms of security.

All in all, following advantages of V2LC are believed to make VLC a strong candidate and as a complementary solution for vehicular communications.

- Low complexity and cost: Due to the much smaller multipath effect, and IM/DD scheme, V2LC transmitter design is easier when compared to heterodyne communication systems. Moreover, already deployed LED lights are demonstrated to be capable of supporting V2LC.
- Scalability: V2LC is foreseen to be utilized as a complementary solution for LoS vehicular communications. Hence, the vehicle density increase in crowded traffic scenarios is not expected to cause interference, contention, or packet collision issues, which may degrade the communication performance.
- Security: Malicious attacks and intentional jamming should be made in LoS distances for V2LC networks. Thus, attacker will be exposed with high possibility.
- Compatibility: Electromagnetic compatibility problem is expected to be minimized with V2LC as RF and visible light occupy different parts of the spectrum.

Spectrum availability: Recently, growing interest to use DSRC allocated spectrum for Wi-Fi is declared due to the scarcity of the RF spectrum. However, with VLC, 10,000 times larger license free spectrum availability is favorable to support multiple V2LC channels simultaneously.

Author details

Bugra Turan[1]* and Seyhan Ucar[2]*

*Address all correspondence to: bturan14@ku.edu.tr; sucar@ku.edu.tr

1 Department of Electrical and Electronics Engineering, Koc University, Istanbul, Turkey

2 Department of Computer Engineering, Koc University, Istanbul, Turkey

References

[1] Digital Light [Internet]. 2017. Available from: https://www.mercedes-benz.com/en/mercedes-benz/next/advanced-engineering/digital-light-headlamps-in-hd-quality [Accessed: February 5, 2017]

[2] ECE Regulation No. 48. Uniform provisions concerning the approval of vehicles with regard to the installation of lighting and light-signaling devices. vol. Rev.1/Add.47/Rev.6. pp. 52-54

[3] Bazzi A, Masini BM, Zanella A, Calisti A. Visible light communications as a complementary technology for the internet of vehicles. Computer Communications. 2016;**93**:39-51

[4] Abdelgader A, Lenan W. The physical layer of the IEEE 802.11 p WAVE Communication Standard: The specifications and challenges. Proceedings of the World Congress on Engineering and Computer Science. 2014;**2**:71

[5] Turan B, Narmanlioglu O, Ergen SC, Uysal M. On the performance of mimo of dm-based intra-vehicular vlc networks. In: 2016 IEEE 84th Vehicular Technology Conference (VTC-Fall); September 18-21, 2016; Montreal, QC, Canada: IEEE; 2016. pp. 1-5

[6] Turan B, Narmanlioglu O, Ergen SC, Uysal M. Physical layer implementation of standard compliant vehicular vlc. In: 2016 IEEE 84th Vehicular Technology Conference (VTC-Fall); September 18-21, 2016; Montreal, QC, Canada: IEEE; 2016. pp. 1-5

[7] Yamazato T, Takai I, Okada H, Fujii T, Yendo T, Arai S, Andoh M, Harada T, Yasutomi K, Kagawa K, et al. Image-sensor-based visible light communication for automotive applications. IEEE Communications Magazine. 2014;**52**(7):88-97

[8] Lee T, Dentai A. Power and modulation bandwidth of gaas-algaas high-radiance led's for optical communication systems. IEEE Journal of Quantum Electronics. 1978;**14**(3):150-159

[9] Zhang H, Yuan Y, Xu W. PAPR reduction for DCO-OFDM visible light communications via semidefinite relaxation. IEEE Photonics Technology Letters. 2014;**26**(17):1718-1721

[10] Yu Z, Baxley RJ, Zhou GT. Iterative clipping for PAPR reduction in visible light OFDM communications. In: Military Communications Conference (MILCOM); October 6-8, 2014; Montreal, QC, Canada: IEEE; 2014. pp. 1681-1686

[11] Turan B, Ucar S, Ergen SC, Ozkasap O. Dual channel visible light communications for enhanced vehicular connectivity. In: Vehicular Networking Conference (VNC), December 16-18, 2015; Kyoto, Japan: IEEE; 2015. pp. 84-87

[12] Uniform Provisions Concerning the Approval of Power-driven Vehicle Front Fog Lamps. 2017. Available from: https://goo.gl/7vXHbK [Accessed: February 1, 2017]

[13] MATRIX LED Lights. 2017. Available from: https://goo.gl/f0OOdi [Accessed: January 20, 2017]

[14] Kim YH, Cahyadi WA, Chung YH. Experimental demonstration of vlc-based vehicle-to-vehicle communications under fog conditions. IEEE Photonics Journal. 2015;**7**(6):1-9

[15] Tseng H-Y, Wei Y-L, Chen A-L, Wu H-P, Hsu H, Tsai H-M. Characterizing link asymmetry in vehicle-to-vehicle visible light communications. In: Vehicular Networking Conference (VNC); December 16-18, 2015; Kyoto, Japan: IEEE; 2015. pp. 88-95

[16] Iwasaki S, Premachandra C, Endo T, Fujii T, Tanimoto M, Kimura Y. Visible light road-to-vehicle communication using high-speed camera. In: Intelligent Vehicles Symposium; June 4-6, 2008; Eindhoven, Netherlands: IEEE; 2008. pp. 13-18

[17] Takai, Isamu, et al.LED and CMOS image sensor based optical wireless communication system for automotive applications. IEEE Photonics Journal. 2013; **5**(5):6801418-6801418

[18] Goto Y, Takai I, Yamazato T, Okada H, Fujii T, Kawahito S, Arai S, Yendo T, Kamakura K. A new automotive vlc system using optical communication image sensor. IEEE Photonics Journal. 2016;**8**(3):1-17

[19] Association et al. IS. Ieee 802.15. 7 standard for local and metropolitan area networks, Part 15.7: Short-Range Wireless Optical Communication Using Visible Light. 2011

[20] Lee SH, Jung S-Y, Kwon JK. Modulation and coding for dimmable visible light communication. IEEE Communications Magazine. 2015;**53**(2):136-143

[21] Amoozadeh M, Deng H, Chuah CN, Zhang HM, Ghosal D. Platoon management with cooperative adaptive cruise control enabled by VANET. Vehicular communications. 2015;**2**(2):110-123

[22] Ishihara S, Rabsatt RV, Gerla M. Improving reliability of platooning control messages using radio and visible light hybrid communication. In: Vehicular Networking Conference (VNC); December 16-18, 2015; Kyoto, Japan: IEEE; 2015

[23] Segata M, Cigno, RL, Tsai HMM, Dressler F. On platooning control using IEEE 802.11 p in conjunction with visible light communications. In: Wireless On-demand Network Systems and Services (WONS), 2016 12th Annual Conference. IEEE; 2016. pp. 1-4

[24] Ucar, S., Ergen, S. C., & Ozkasap, O. Security vulnerabilities of IEEE 802.11 p and visible light communication based platoon. In *Vehicular Networking Conference (VNC), 2016, IEEE* (pp. 1-4). IEEE

[25] Viriyasitavat W, Yu S-H, Tsai H-M. Short paper: Channel model for visible light communications using off-the-shelf scooter taillight. In: Vehicular Networking Conference (VNC); December 16-18, 2013; Boston, MA, USA: IEEE; 2013

[26] Luo P, Ghassemlooy Z, Le Minh H, Bentley E, Burton A, Tang X. Performance analysis of a car-to-car visible light communication system. Applied Optics. 2015;**54**(7):1696-1706

[27] Ucar S, Turan B, Coleri Ergen S, Ozkasap O, Ergen M. Dimming support for visible light communication in intelligent transportation and traffic system. In: Urban Mobility and Intelligent Transportation System (UMITS), IEEE/IFIP Network Operations and Management Symposium (NOMS); April 25-29, 2016; Istanbul, Turkey: IEEE; 2016

[28] Abualhoul MY, Marouf M, Shagdar O, Nashashibi F. Platooning control using visible light communications: A feasibility study. In: Intelligent Transportation Systems-(ITSC), 2013 16th International IEEE Conference. IEEE. 2013. pp 1535-1540

[29] Rajagopal S, Roberts RD, Lim SK. IEEE 802.15.7 visible light communication: modulation schemes and dimming support. IEEE Communications Magazine. 2012;**50**(3):72-82

[30] Wu L, Zhang Z, Dang J, Liu H. Adaptive modulation schemes for visible light communications. Journal of Lightwave Technology. 2015;**33**(1):117-125

[31] Wang M, Wu J, Yu W, Wang H, Li J, Shi J, Luo C. Efficient coding modulation and seamless rate adaptation for visible light communications. IEEE Wireless Communications. 2015;**22**(2):86-93

[32] Checkoway S, McCoy D, Kantor B, Anderson D, Shacham H, Savage S, Koscher K, Czeskis A, Roesner F, Kohno T. Comprehensive experimental analyses of automotive attack surfaces. In: Proceedings of the 20th USENIX Conference on Security; August 8-12, 2011; San Francisco, CA: SEC, USENIX Association; 2011

[33] Miller C, Valasek C. Adventures in automotive networks and control units. DEF CON 21 (2013):260-264

[34] Miller C, Valasek C. A survey of remote automotive attack surfaces. Black Hat USA. 2014

[35] Miller C, Valasek C. Remote exploitation of an unaltered passenger vehicle. Black Hat USA. 2015

[36] Sun J, Zhang C, Zhang Y, Fang Y. An identity-based security system for user privacy in vehicular ad hoc networks. IEEE Transactions on Parallel and Distributed Systems. 2010;**21**(9):1227-1239

[37] Xi Y, Sha K, Shi W, Schwiebert L, Zhang T. Enforcing privacy using symmetric random Key-Set in vehicular networks. In: IEEE Eighth International Symposium on Autonomous Decentralized Systems (ISADS); March 21-23, 2007; Sedona, AZ, USA: IEEE; 2007

[38] Mejri MN, Achir N, Hamdi M. A new group Diffie-Hellman key generation proposal for secure VANET communications. In: 13th IEEE Annual Consumer Communications Networking Conference (CCNC); January 9-12, 2016; Las Vegas, NV, USA: IEEE; 2016

[39] Li-1st. Available from: http://purelifi.com/li-fire/li-1st/

[40] Google Keyczar Tool. Available from: https://goo.gl/LMuY1l

[41] pureLifi. Available from: http://purelifi.com

[42] LEDFog101 OSRAM. Available from: http://goo.gl/ty5zEC

[43] Vishay Infra-red (IR) Diodes. Available from: http://goo.gl/2DIKcq

[44] VEhicular NeTwork Open Simulator (VENTOS). Available from: http://goo.gl/OueFkO